Abrasion Resistance of Materials

Abrasion Resistance of Materials

Editor

Leroy Sidney

scitus
academics

Abrasion Resistance of Materials
Edited by **Leroy Sidney**

ISBN: 978-1-68117-236-1
Library of Congress Control Number: 2016934768

SCITUS Academics LLC,
www.scitusacademics.com
Box No. 4766, 616 Corporate Way,
Suite 2, Valley Cottage,
NY 10989

Notice

Preface

Abrasion is the process of scuffing, scratching, wearing down, marring, or rubbing away. It can be intentionally imposed in a controlled process using an abrasive. Abrasion can be an undesirable effect of exposure to normal use or exposure to the elements. Abrasion resistance is a property which allows a material to resist wear. Materials which are abrasion resistant are useful for situations in which mechanical wearing and damage can occur, including delicate applications such as the construction of space shuttle components. Numerous abrasion resistant products are employed for a variety of applications, including products which can be custom fabricated to meet the needs of specific users. When a product has abrasion resistance, it will resist erosion caused by scraping, rubbing, and other types of mechanical wear. This allows the material to retain its integrity and hold its form. This can be important when the form of a material is critical to its function, as seen when moving parts are carefully machined for maximum efficiency. Abrasion resistant materials can be used for both moving and fixed parts in settings where wearing could become an issue.The resistance of materials and structures to abrasion can be measured by a variety of test methods. These often use a specified abrasive or other controlled means of abrasion.There are a number of ways to make a material resistant to abrasion. One option is to utilize a special coating which creates a hardened layer over the material and resists friction. Some materials are also naturally extremely hard, and are ideal for settings in which abrasion resistance will be desirable. Other materials can be specifically formulated for increased hardness, as seen in plastics facilities which manufacture abrasion resistant plastics with the use of chemicals which harden and strengthen the plastic. Abrasion Resistance of Materials covers on the abrasion characteristics of various materials such as polymer nanocomposites, high performance fabrics, textiles, etc.

Table of Contents

Chapter 1

An Investigation on the Wear Resistance and Fatigue Behaviour of Ti-6Al-4V Notched Members Coated with Hydroxyapatite Coatings

Reza H Oskouei[1,2,], Khosro Fallahnezhad[2] and Sushmitha Kuppusami[2]*

[1] *Medical Device Research Institute, School of Computer Science, Engineering and Mathematics, Flinders University, Tonsley 5042, Australia*
[2] *Discipline of Mechanical Engineering, School of Computer Science, Engineering and Mathematics, Flinders University, Tonsley 5042, Australia*

ABSTRACT

In this study, surface properties of Ti-6Al-4V alloy coated with hydroxyapatite coatings were investigated. Wear resistance and fatigue behaviour of samples with coating thicknesses of 10 and 50 μm as well as uncoated samples were examined. Wear experiments demonstrated that the friction factor of the uncoated titanium decreased from 0.31 to 0.06, through a fluctuating trend, after 50 cycles of wear tests. However, the friction factor of both the coated samples (10 and 50 μm) gradually decreased from 0.20 to 0.12 after 50 cycles. At the end of the 50th cycle, the penetration depth of the 10 and 50 μm coated samples were 7.69 and 6.06 μm, respectively. Fatigue tests showed that hydroxyapatite coatings could improve fatigue life of a notched Ti-6Al-4V member in both low and high cycle fatigue zones. It was understood, from fractography of the fracture surfaces, that the fatigue zone of the uncoated specimens was generally smaller in comparison with that of the coated specimens. No significant difference was observed between the fatigue life of coated specimens with 10 and 50 μm thicknesses.

Keywords: titanium alloys; hydroxyapatite coatings; wear; fatigue

1. INTRODUCTION

Owing to their excellent mechanical properties and biocompatibility, metallic biomaterials are largely used in different medical applications such as orthopaedic implants, bone plates, dental implants and cardiovascular devices [1,2,3]. It is reported that 70%–80% of implants used for biomedical applications are made of metals and metallic alloys [4] for which high strength, low elastic modulus, excellent wear and corrosion resistance, and good biocompatibility are of paramount importance [5]. Due to the exposure to the corrosive environment of the body, metallic implants may suffer from corrosion which can be intensified by the effect of wear induced by coming into contact with a hard surface [6]. This can result in toxicity, adverse local tissue reaction and ultimately failure of the metallic implant [7,8]. Given the cyclic nature of the mechanical loads applied during physical activities, fatigue behaviour of implants is also important.

Surface modification methods, particularly coatings, have been extensively used to improve mechanical and surface properties of metals in different applications [9,10,11,12]. Wang and Zreiqat [3] reviewed a wide range of surface coatings used for metallic biomaterials along with their deposition methods and properties including corrosion, wear, biocompatibility and antibacterial activity. Hydroxyapatite (HA) coatings [$Ca_{10}(PO_4)_6(OH)_2$] are widely used in prosthetic applications as they possess a similar chemical composition and crystal unit to apatite of the bone [13,14]. It favourably accelerates the bone formation process as it contains calcium phosphate [15] which makes it suitable for bone substitution and reconstruction [10,15,16,17].

HA coatings have shown success in implant fixation (*in-vivo*) which enhances the implant life by increasing the strength and rate of implant integration [18]. The dense layer of coating onto the surface of implant material is mainly for bio-integration and increased implant stability to the bone tissue [19,20]. In addition to its bioactive role, HA coatings can minimise ion release of metals in the physiological environment of the body [15].

However, HA coatings have a brittle nature and low strength under wear conditions [21,22]. Balani *et al.* [23] evaluated wear behaviour of hydroxyapatite coatings reinforced by plasma-sprayed carbon nanotubes in a simulated body fluid environment. Their results demonstrated considerable improvements in the fracture toughness (56%) and crystallinity (27%). Moreover, coated implants may suffer from fatigue damage under cyclic loading in the body. Thus, fatigue response of coated metallic biomaterials needs to be investigated. The effect of surface coatings on the fatigue behaviour of biomaterials, especially metals, has been investigated previously. Apachitei *et al.* [24] investigated fatigue behaviour of Ti-6Al-4V and Ti-6Al-7Nb alloys coated with plasma electrolytic oxidation coatings. Their study showed that this type of coatings reduces fatigue resistance of these alloys. Vadiraj and Kamaraj [25] studied the effect of titanium nitride coatings on the fretting wear properties of titanium alloys. They concluded that physical vapour deposition (PVD) TiN coatings can improve fretting fatigue properties of titanium alloys.

There have been however limited studies on the fatigue behaviour of metallic biomaterials coated with HA coatings. Lynn*et al.* [26] investigated the effect of HA coating thickness on the fatigue behaviour of Ti-6Al-4V. They found that changing the thickness of atmospheric plasma-spray (APS) HA coatings from 0 to 100 μm does not affect the fatigue life of the alloy, but thickness of 150 μm considerably reduces the fatigue resistance of Ti-6Al-4V. They also investigated the effect of heat treatment on the fatigue behaviour of this alloy coated with HA coatings [27]. It was shown that a heat treatment at 400 °C for 90 h significantly reduces the fatigue resistance of this alloy. Fatigue tests in these studies were limited to one constant stress amplitude (620 MPa); whereas, stress *versus* number of cycles (S-N) curves could be more useful in understanding the overall fatigue behaviour. Also, implants usually have complex geometries with fillets and reduced sections which can act as a notch. Hence, fatigue behaviour of a notched implant material coated with HA coatings should be studied.

In this study, surface properties of hydroxyapatite coatings deposited onto Ti-6Al-4V substrate is characterised. Two thickness of 10 and 50 μm were studied. Wear tests were conducted to evaluate wear resistance of the HA coating layer with two different thicknesses. The effect of HA coatings thickness on the fatigue behaviour of Ti-6Al-4V notched specimens was also investigated in different cyclic load levels. Fatigue crack initiations and propagations were examined by scanning electron microscopy (SEM) to identify crack origins and detect coating delaminations after failure.

2. EXPERIMENTAL SECTION

2.1. Sample Preparation

Titanium alloys are one of the most commonly used metals for biomedical applications because they offer superior properties which include low density, moderate elastic modulus (110 GPa), good corrosion resistance and high strength [2]. In this work, Ti-6Al-4V alloy, ELI (Extra Low Interstitials) grade (Grade 23) was used as substrate. Ti-6Al-4V ELI contains reduced amounts of oxygen, nitrogen, carbon and iron. In this alloy, lower interstitials can improve ductility and offer better fracture toughness. The alloy was purchased as a 0.50-inch-diameter bar from Magellan Metals, South Norwalk, CT, USA. Small disc samples were cut from the bar using Secotom-50 (Struers, Ballerup, Denmark). The samples had the same diameter as the bar (12.7 mm) and a 5-mm-thickness.

After the cutting process, one face of the discs was mirror polished using Tegramin-25 (Struers, Ballerup, Denmark) to remove surface scratches.

Three batches of samples were prepared. These included uncoated titanium samples, HA coated samples with a coating thickness of 5–15 μm, and HA coated samples with a coating thickness of 45–55 μm. Hydroxyapatite coatings were deposited onto the titanium surface using a thermal plasma spray process by Himed, Old Bethpage, NY, USA. All the samples were coated in the targeted thickness ranges. Nominal thicknesses of 10 and 50 μm were used in this study for the afore-mentioned thickness ranges, respectively. Prior to applying the HA coating, the disc samples were degreased using 3% Alconox, distilled water and 99% alcohol. The polished surface was then grit-blasted with apatitic abrasive (MCD, 180–300 μm, Himed, Old Bethpage, NY, USA) and passivated as per ASTM F86-04 [28]. The coated sample is shown in Figure 1a. Fatigue test specimens with a reduced section in diameter, as shown in Figure 1b, were manufactured from the Ti-6Al-4V bar. Similar to the disc samples, fatigue test specimens were coated with hydroxyapatite coatings with two thicknesses of 10 and 50 μm. The notched section was only coated where final fracture was expected to occur (Figure 1c).

(a)

(b)

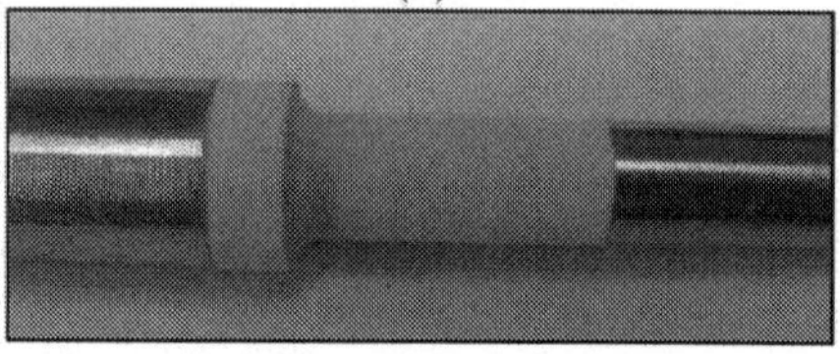

(c)

Figure 1. (**a**) Ti-6Al-4V disc sample coated with hydroxyapatite coatings; (**b**) fatigue test specimen, dimensions in mm; and (**c**) notched section of the fatigue test specimen coated with hydroxyapatite coatings.

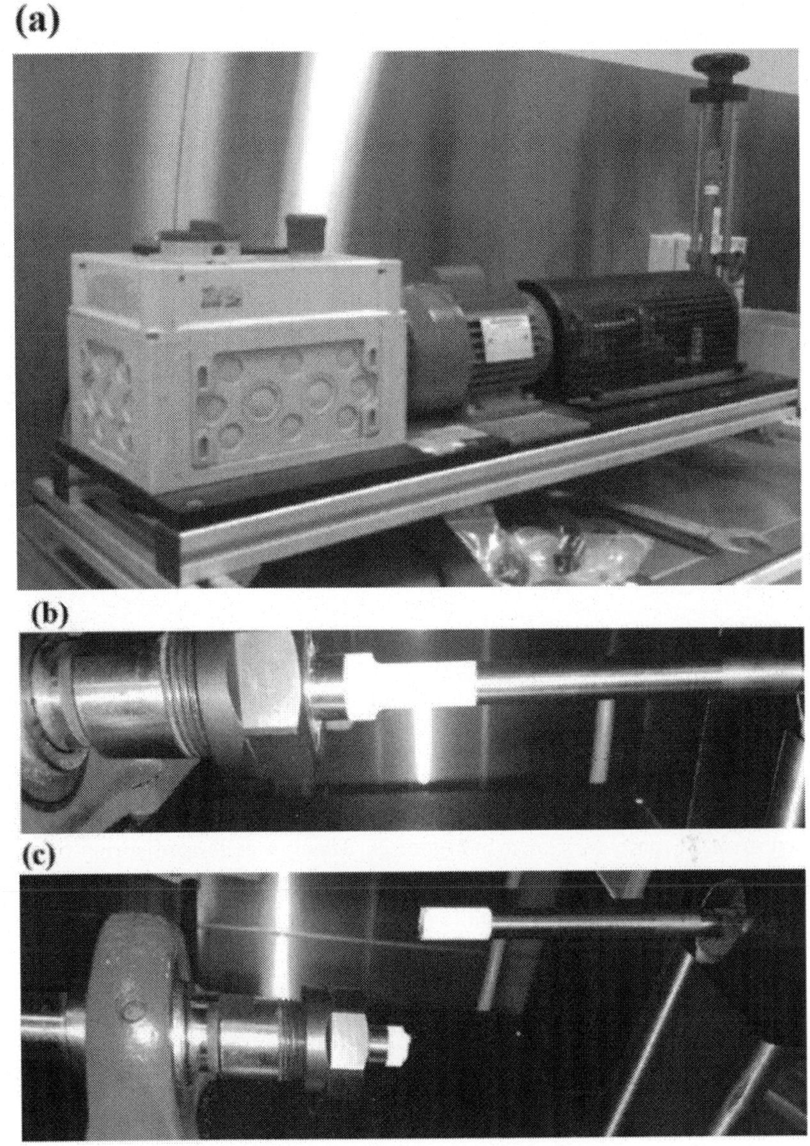

Figure 2. (**a**) Fatigue testing machine, GUNT WP 140; (**b**) HA coated specimen under fatigue loading; and (**c**) HA coated specimen after fatigue failure.

2.2. Coating Characterisations

The coated disc samples were used to characterise surface properties of the coating. Scanning electron microscopy, SEM, (FEI, Inspect F50, Hillsboro, OR, USA) and energy dispersive spectroscopy (EDS) were used to examine the microstructure and chemical composition of the coating and Ti-6Al-4V substrate. In addition, surface topography and roughness of the coating and titanium substrate were characterised using a surface profilometer (WYKO NT9100, Veeco, Plainview, NY, USA) to determine the average roughness parameters of the surface.

2.3. Wear Assessments

Wear resistance of the HA coated samples and uncoated titanium sample was examined using a scratch tester mounted on an IBIS nano-indentation system (M/S Fisher-Cripps Laboratories Pvt. Limited, Killarney Heights, NSW, Australia). This technique was used to evaluate wear resistance of the material by applying a certain normal force to the surface along with a displacement amplitude for a number of passes. A 200 μm diamond sphero-cone tip was used with a normal force of 50 mN and a displacement amplitude of 100 μm (in both left and right directions) with a velocity of 5 μm/s over the surface. The coefficient of friction and penetration depth were measured and recorded during the test. The wear test was performed for 50 passes for each sample and the data were recorded in every two passes.

2.4. Fatigue Tests

Fatigue tests were conducted to study the effect of HA coatings on the fatigue resistance of Ti-6Al-4V alloy with a notch, as shown in Figure 1b. To accomplish this, a rotating fatigue testing machine (G.U.N.T. Gerätebau GmbH, Hamburg, Germany, model WP 140) was used to apply completely reversed bending stresses (stress ratio of $R = -1$) with a frequency of 46.7 Hz (spindle speed of 2800 rpm), as shown in Figure 2a. Using a calibrated spring balance together with a floating bearing, the specimens were loaded to desired levels. Depending on the load level, bending stresses applied to the top and bottom faces of the specimens were between 450 and 600 MPa. This range of stress amplitudes was found to be reasonable to address both low cycle and high cycle fatigue behaviours for the specimens. Fatigue life was considered as the complete separation of the specimens (Figure 2c). For each type of specimen and stress amplitude, three fatigue tests were performed and the average fatigue life was used in S-N curves.

3. RESULTS AND DISCUSSION

3.1. Coating Characteristics

SEM images of both the HA coated samples (with 10 and 50 μm thicknesses) and the uncoated titanium sample are shown in Figure 3. The image of both thicknesses demonstrate that the hydroxyapatite coatings are deposited uniformly onto the titanium substrate surface. It is apparent that in spite of some micro-voids, coatings are tightly adhered together. The mirror polished surface of the uncoated Ti-6Al-4V can be seen in Figure 3c before applying the HA coatings.

EDS results of the coating layer are given in Figure 4. It is noted that the EDS results were the same for both thicknesses of the coating layer, as expected. The elemental analysis confirmed the presence of calcium (55%), phosphorus (20%) and oxygen (25%), all in percentage weight (wt%), in hydroxyapatite coatings used in this work.

The surface roughness parameters for uncoated, 10 μm coated and 50 μm coated samples were recorded in five different areas over the sample surface. The arithmetic average (R_a), root mean squared (R_q) and maximum vertical distance between the highest and lowest data points (R_{max}) were determined, as given in Table 1. As can be seen, there is no considerable difference in the surface roughness measures of the two coated samples indicating that the roughness of the 10 and 50 μm coated samples is very similar. This confirms that the surface roughness, as a component of the surface texture, was independent of the coating thickness. However, the surface roughness of the uncoated titanium sample was much smaller (R_a = 15.49 nm against R_a = 2.60–2.66 μm) which was obviously because of the highly polished surface finish of the substrate. The 3D surface profile for one of the five scanned areas of these samples is shown in Figure 5 where the similarity between the two coated samples can be seen in terms of the surface profile.

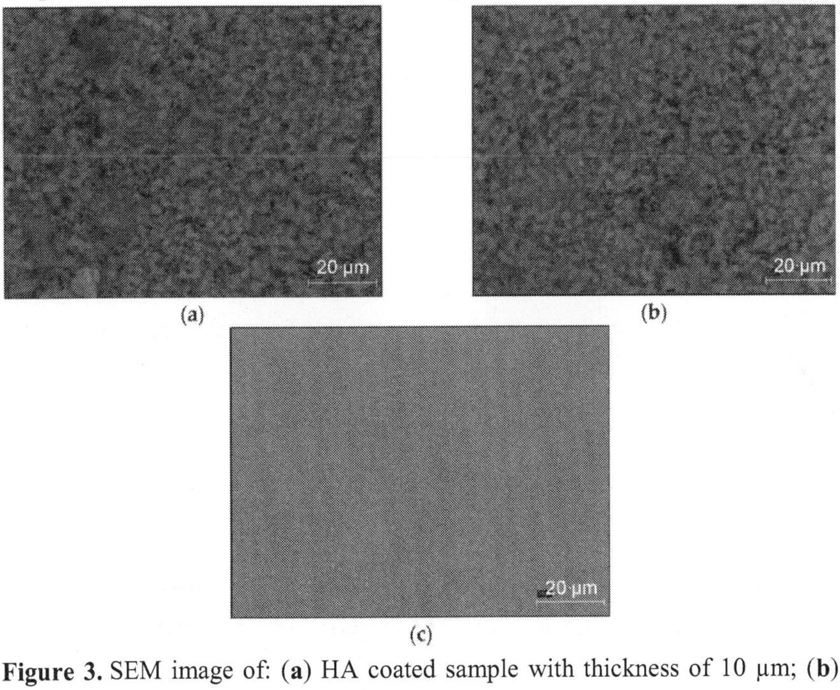

Figure 3. SEM image of: (**a**) HA coated sample with thickness of 10 μm; (**b**) HA coated sample with thickness of 50 μm; and (**c**) uncoated Ti-6Al-4V sample.

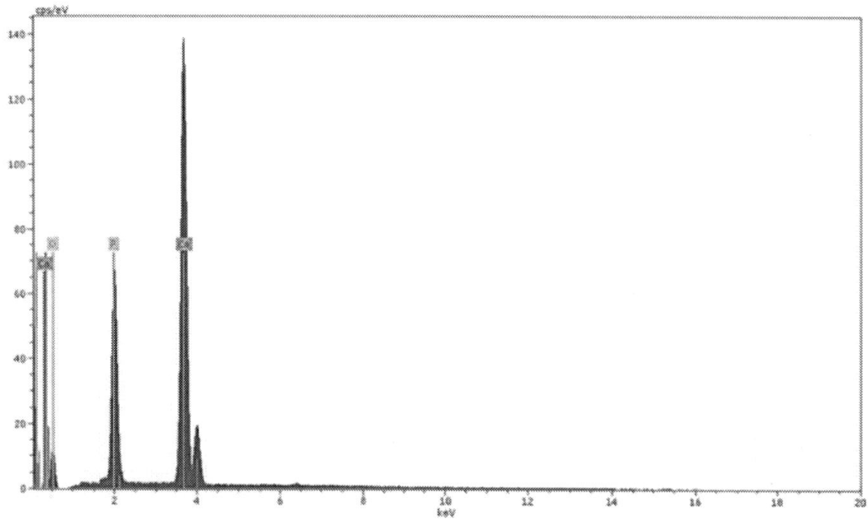

Figure 4. EDS results of hydroxyapatite coatings used in this study.

Table 1. Surface roughness measures (R_a, R_q and R_{max}) of different areas for three batches of samples.

Tested Areas	10 μm Coated Sample (μm)	50 μm Coated Sample (μm)	Uncoated Sample (nm)
Area 1	R_a = 2.46 R_q = 3.15 R_{max} = 12.4	R_a = 2.60 R_q = 3.19 R_{max} = 13.00	R_a = 19.18 R_q = 23.39 R_{max} = 10.10
Area 2	R_a = 2.47 R_q = 3.16 R_{max} = 12.20	R_a = 3.14 R_q = 3.84 R_{max} = 13.00	R_a = 13.99 R_q = 17.79 R_{max} = 9.20
Area 3	R_a = 2.90 R_q = 3.68 R_{max} = 14.30	R_a = 2.42 R_q = 3.04 R_{max} = 11.62	R_a = 16.82 R_q = 22.15 R_{max} = 8.76
Area 4	R_a = 2.21 R_q = 2.90 R_{max} = 13.50	R_a = 2.67 R_q = 3.40 R_{max} = 12.80	R_a = 13.11 R_q = 16.61 R_{max} = 8.72
Area 5	R_a = 2.97 R_q = 3.78 R_{max} = 13.90	R_a = 2.47 R_q = 3.02 R_{max} = 11.00	R_a = 14.35 R_q = 18.13 R_{max} = 9.14
Average	R_a = 2.60 R_q = 3.33 R_{max} = 13.26	R_a = 2.66 R_q = 3.30 R_{max} = 12.28	R_a = 15.49 R_q = 19.61 R_{max} = 9.18

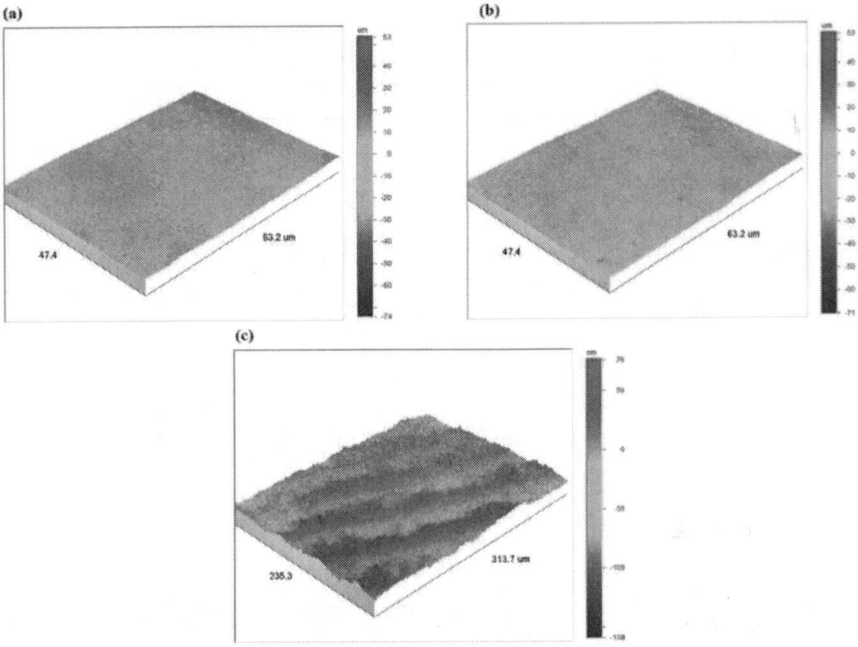

Figure 5. 3D surface profile of: (**a**) HA coated sample with thickness of 10 μm; (**b**) HA coated sample with thickness of 50 μm; and (**c**) uncoated Ti-6Al-4V sample.

3.2. Wear Resistance Results

Friction behaviour of the three samples is shown in Figure 6a in the form of friction factor through 50 passes of the sliding test. It is apparent that the friction factor of the uncoated titanium sample was considerably greater than that of the coated samples with both thicknesses at the beginning of the test. The friction factor was approximately steady by around 18 passes after which, there was an overall reduction in the friction factor of the uncoated sample. The trend remained constant again from the 34th to the 50th pass of the test. Variations in friction factor were more uniform for both the coated samples. At the start of the test, the friction factor for these two samples was around 0.20. With an increase in the number of passes, the friction factor gradually decreased to 0.12. The interesting point is that at the beginning of the test, the uncoated sample had a greater friction factor in comparison with the coated samples. However, after 50 passes, the friction factor of the uncoated sample was found to be smaller than the coated samples. Such behaviours indicate that friction factor varies with the number of passes of the sliding test. Figure 6b shows the behaviour of the penetration depth *versus* the number of passes from the wear experiments on the three samples. It can be seen from the plots that the penetration depth for the uncoated sample was almost constant (0.74 μm) during the wear test; whereas, the penetration depth of the coated samples gradually increased during the test. The penetration depth of the 10 μm coating was greater than that of the 50 μm coating. The maximum penetration depths for the 10 and 50 μm coated samples were 7.69 and 6.06 μm, respectively. Given the maximum penetration depths were still within the thickness range of both the coated samples (10 and 50 μm), the friction factors in both the coated samples, as shown in Figure 6a, were found to be similar during the wear experiments.

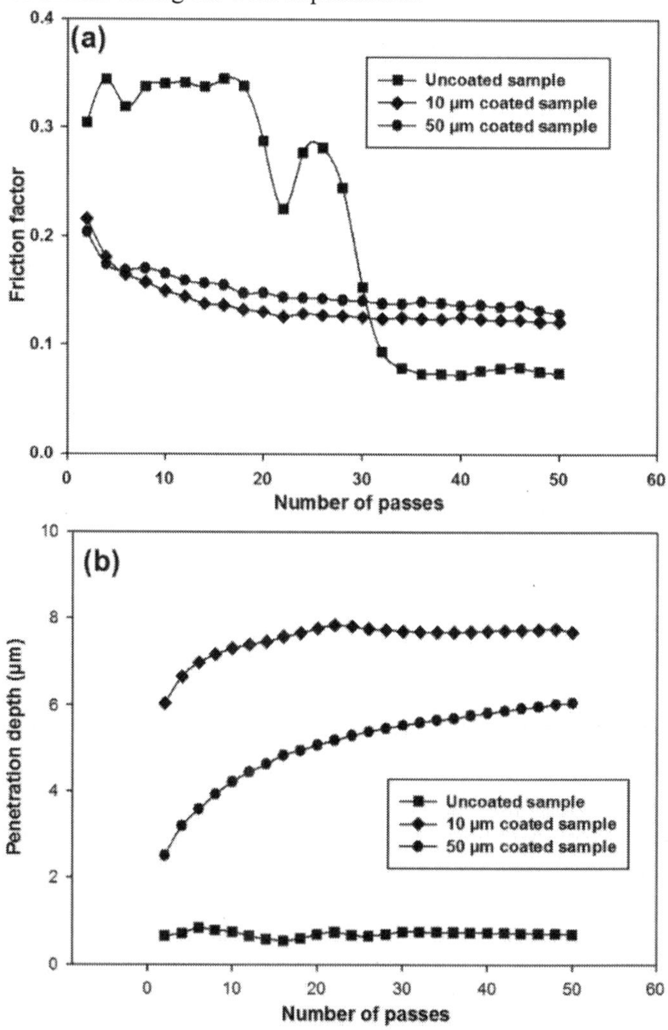

Figure 6. Wear test results of: (**a**) friction factor *versus* number of passes; (**b**) depth of penetration versus number of passes for three batches of samples.

3.3. Fatigue Test Results

Figure 7 presents the S-N curves of the three batches of tested specimens. In general, HA coatings were found to improve the fatigue resistance of Ti-6Al-4V alloy, especially at high cycle fatigue zone. It can be seen that under stress amplitudes between 520 and 590 MPa, 10 μm coated specimen experienced the highest number of cycles before failure. For instance, under 590 MPa, the 10 and 50 μm coated samples failed after 139,000 and 96,000 cycles, respectively. However, the uncoated specimen had 58,000 cycles. At 6 million cycles, two red points are marked on the curves for the uncoated and coated specimens. These two points were considered as the fatigue limit that were 468 and 511 MPa for the uncoated and both coated specimens, respectively.

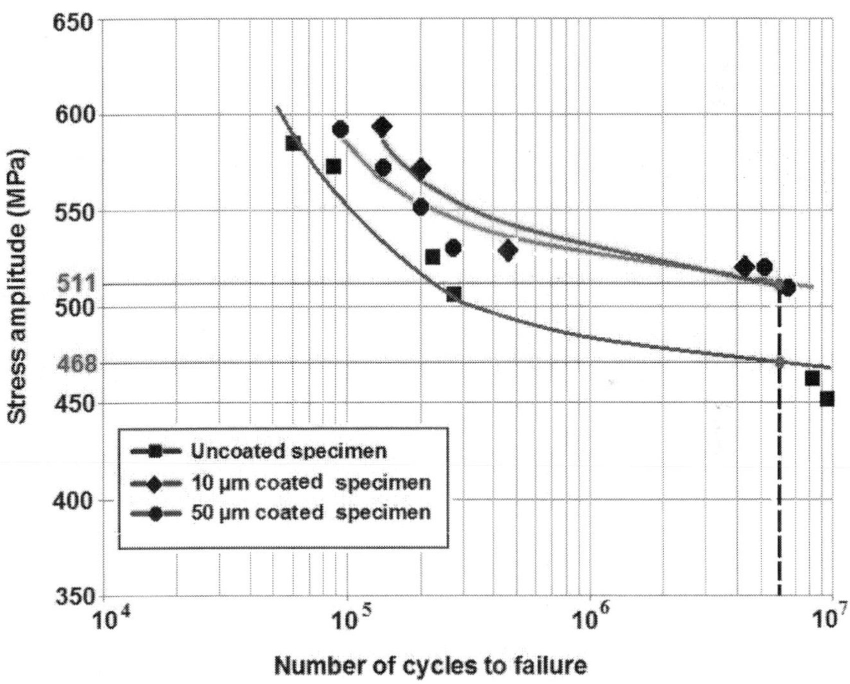

Figure 7. S-N curves of uncoated Ti-6Al-4V, 10 μm HA coated and 50 μm HA coated specimens, $R = -1$.

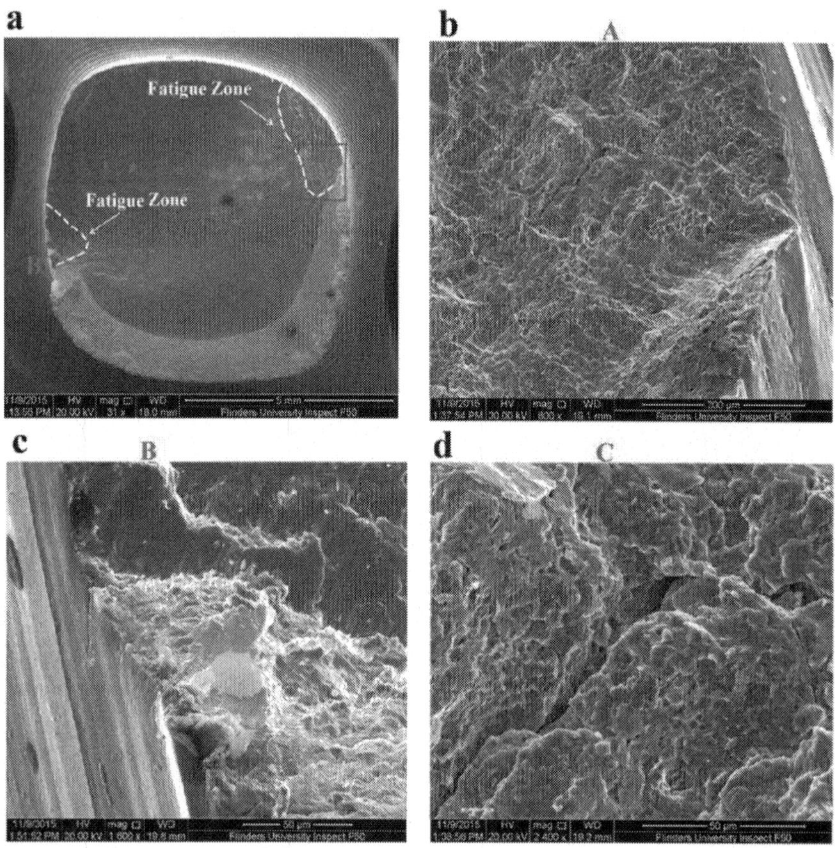

Figure 8. SEM images of fatigue fracture surface of uncoated Ti-6Al-4V specimen failed under stress amplitude of 570 MPa and $R = -1$: (**a**) fracture surface showing fatigue zones; (**b**) fatigue crack initiation site A; (**c**) fatigue crack initiation site B; and (**d**) higher magnification image of a deep crack in area C.

Fatigue fracture surfaces of 10 and 50 μm coated specimens and uncoated specimen were investigated using SEM to evaluate fatigue crack initiations and propagations along with their fatigue zone over the fracture surface. Furthermore, it was important to evaluate the coating adherence to the substrate after fatigue failure. SEM images of three specimens including uncoated titanium specimen, 10 μm HA coated and 50 μm HA coated specimens failed at stress amplitude of 570 MPa are shown in Figure 8, Figure 9 and Figure 10, respectively. The figures present a low magnification image of the whole fracture surface in which fatigue zones are indicated. There are also higher magnification images showing fatigue crack initiation sites and crack propagations towards the specimen centre. Due to the presence of the notch, and as a consequence, the stress concentration all around the specimen circumference, cracks initiated from the edge of the notch, as expected. The interesting finding was that both the 10 and 50 μm coated specimens demonstrated the presence of the coating layer at the edge of the fracture surface which confirms a good adherence between the HA coating and substrate under cyclic loading (Figure 9b and Figure 10b).

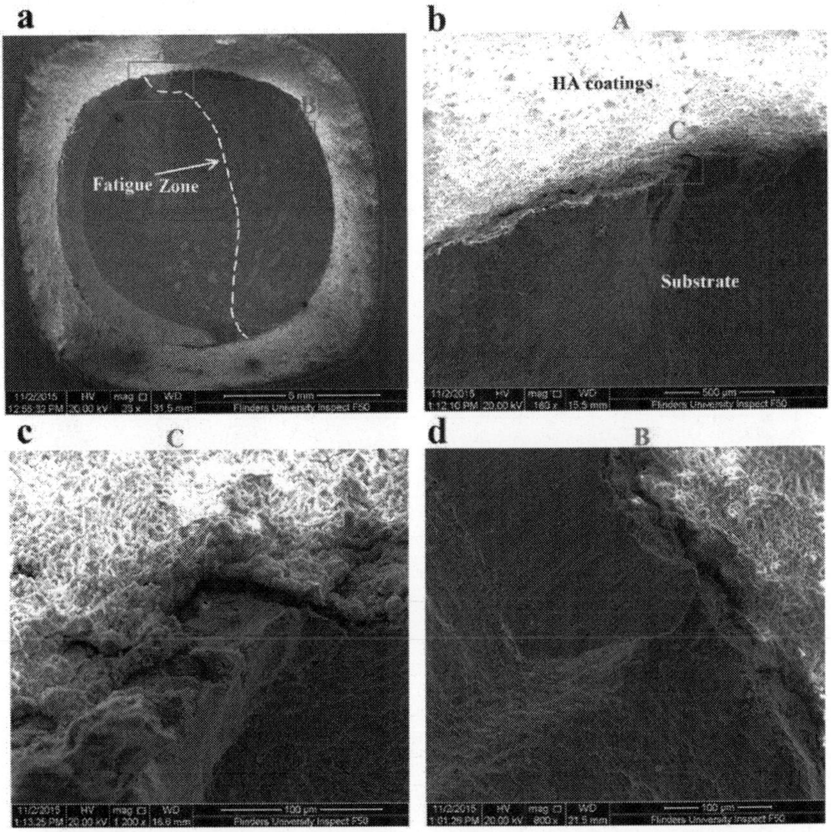

Figure 9. SEM images of fatigue fracture surface of 10 μm HA coated specimen failed under stress amplitude of 570 MPa and $R = -1$: (**a**) fracture surface showing fatigue zone; (**b**) fatigue crack initiation site A; (**c**) higher magnification image of fatigue crack initiation point C; and (**d**) higher magnification image of fatigue crack initiation site B.

An investigation of the SEM images also showed that the fatigue zone of the uncoated titanium specimen was generally smaller in comparison with the coated specimens. It can be concluded that the uncoated specimen had less number of cycles for the crack propagation stage and thus smaller fatigue zones were formed over the fracture surface. In other words, under the same stress amplitude, coated specimens tolerated more cycles for the crack propagation stage resulting in an improvement in the fatigue life, as can be seen also in the S-N curves of Figure 7. Moreover, the fatigue experiments showed no significant differences between the 10 and 50 μm coated specimens. The SEM images of these coated specimens showed similar fatigue fracture features mainly in terms of the presence of multiple crack initiations and amount of area that is influenced by crack propagations. This together with the fact that all the crack initiations occurred at the interface of the titanium substrate and the coating layer could explain the similar fatigue behaviour of both the 10 and 50 μm coated specimens. Final fracture zone in all the three specimens showed ductile fracture features. Figure 11 demonstrates the final fracture zone of the 10 μm coated specimen in which dimples can be seen over the surface.

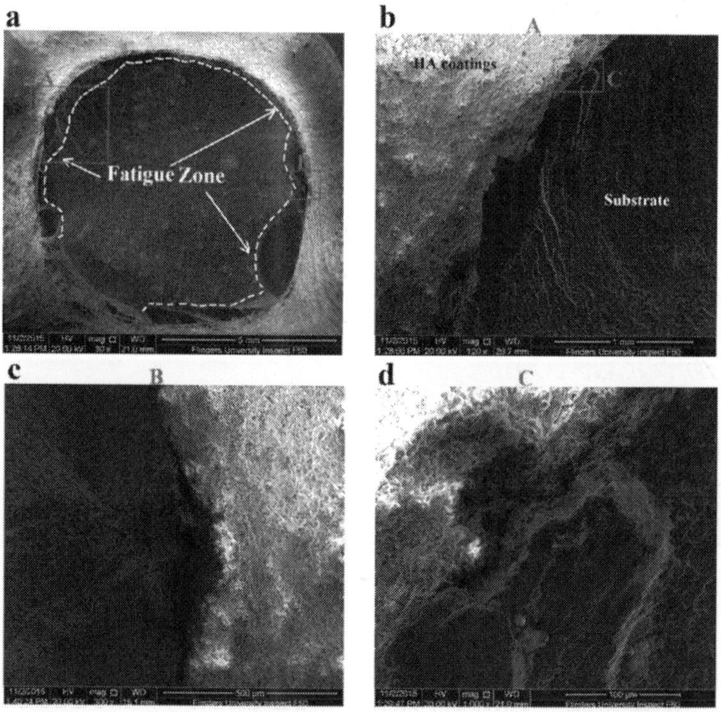

Figure 10. SEM images of fatigue fracture surface of 50 μm HA coated specimen failed under stress amplitude of 570 MPa and $R = -1$: (**a**) fracture surface showing fatigue zones; (**b**) fatigue crack initiation site A; (**c**) fatigue crack initiation site B; and (**d**) higher magnification image of fatigue crack initiation point C.

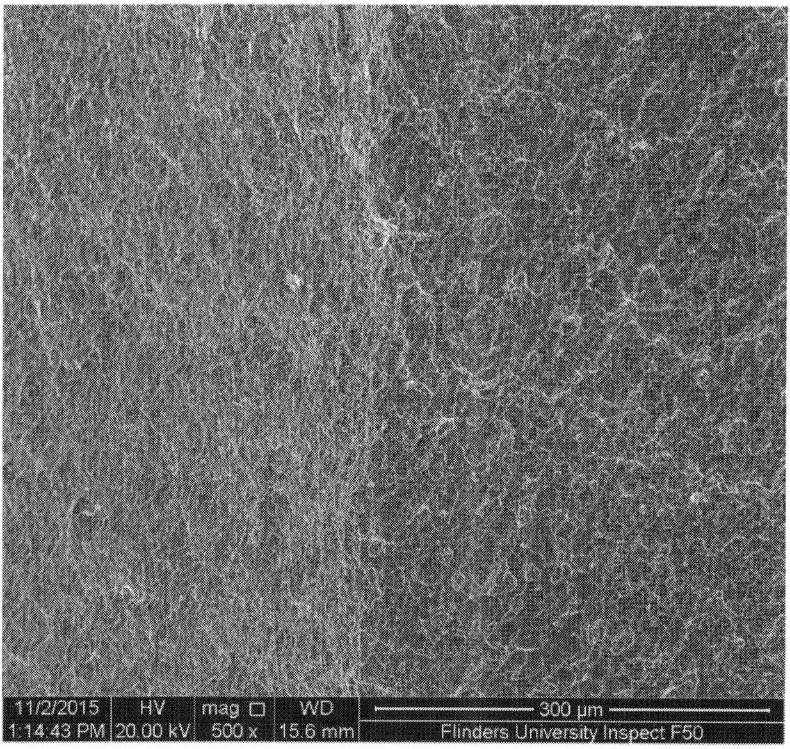

Figure 11. SEM image of the final fracture zone, taken from the 10 µm HA coated specimen.

4. CONCLUSIONS

Surface properties of Ti-6Al-4V substrate coated with hydroxyapatite coatings were characterised. SEM images of the coated samples demonstrated that HA coatings were uniformly deposited onto the substrate surface. The surface roughness measures (R_a, R_q *and* R_{max}) for coated samples with 10 and 50 µm thicknesses were found to be similar. However, roughness parameters were much smaller in the mirror-polished uncoated titanium sample. The friction factor of both the coated samples was also similar decreasing from 0.20 to 0.12 after 50 cycles of the sliding test. This similarity was because the maximum penetration depths of the wear experiments were well within the thickness ranges of the coating. Although at the start of the test the friction factor of the uncoated sample was greater than the coated samples, after 50 cycles, the friction factor became less than that of the coated samples. Fatigue resistance of a notched specimen of Ti-6Al-4V coated with hydroxyapatite coatings was also studied. S-N curves of the coated and uncoated specimens were presented. It was found that HA coatings could improve fatigue life of the titanium notched specimen in both low and high cyclic fatigue zones. SEM images of the fracture surface showed smaller fatigue zones in the uncoated specimen in comparison with the coated ones. The coating layer was found to remain adhered to the substrate under fatigue loading. Moreover, the fatigue behaviour of both the 10 and 50 μm coated specimens was found to be similar. This was justified based on the similarity between the fatigue fracture features that was observed in both the coated specimens.

ACKNOWLEDGMENTS

The authors would like to thank Himed, Old Bethpage, NY, USA for coating titanium samples with hydroxyapatite coatings, Peter Abigail from Technical and Further Education (TAFE), Regency Park, SA, Australia for providing access to the fatigue testing machine, Deepali Deepali for performing fatigue tests and Ari Mouratidis from Struers Australia for polishing titanium samples. Also, the authors would like to acknowledge the Australian National Fabrication Facility (ANFF), Ian Wark Research Institute (University of South Australia) for providing access to Surface Profilometer.

AUTHOR CONTRIBUTIONS

Reza Oskouei designed all the experiments in this study including surface characterisations, wear & friction, SEM/EDS and fatigue. He managed and supervised all the project tasks including experimental tests and analysis of the results and contributed to the SEM and EDS experiments as well as write up the manuscript. Khosro Fallahnezhad illustrated the graphs and discussed the results and contributed to the writing of the manuscript. He also contributed to the SEM and EDS experiments. Sushmitha Kuppusami assisted with performing surface characterisation tests. She also partly contributed to the writing of the manuscript.

REFERENCES

1. Niinomi, M. Metallic biomaterials. *J. Artif. Organs* **2008**, *11*, 105–110.
2. Navarro, M.; Michiardi, A.; Castaño, O.; Planell, J. Biomaterials in orthopaedics. *J. R. Soc. Interface* **2008**, *5*, 1137–1158.
3. Wang, G.; Zreiqat, H. Functional coatings or films for hard-tissue applications. *Materials* **2010**, *3*, 3994–4050.
4. Li, Y.; Yang, C.; Zhao, H.; Qu, S.; Li, X.; Li, Y. New developments of ti-based alloys for biomedical applications.*Materials* **2014**, *7*, 1709–1800.

5. Temenoff, J.; Mikos, A.G. *Biomaterials: The Intersection of Biology and Materials Science*, 1st ed.; Prentice Hall: Upper Saddle River, NJ, USA, 2008; pp. 235–258.
6. Hussenbocus, S.; Kosuge, D.; Solomon, L.B.; Howie, D.W.; Oskouei, R.H. Head-neck taper corrosion in hip arthroplasty. *BioMed Res. Int.* **2015**, *2015*.
7. Hanawa, T. Recent development of new alloys for biomedical use. *Mater. Sci. Forum* **2006**, *512*, 243–248.
8. Farhoudi, H.; Oskouei, R.H.; Jones, C.F.; Taylor, M. A novel analytical approach for determining the frictional moments and torques acting on modular femoral components in total hip replacements. *J. Biomech.* **2015**, *48*, 976–983.
9. Oskouei, R.H.; Ibrahim, R.N. Improving fretting fatigue behaviour of Al 7075-T6 bolted plates using electroless Ni–P coatings. *Int. J. Fatigue* **2012**, *44*, 157–167.
10. Mohseni, E.; Zalnezhad, E.; Bushroa, A.R. Comparative investigation on the adhesion of hydroxyapatite coating on Ti-6Al-4V implant: A review paper. *Int. J. Adhes. Adhes.* **2014**, *48*, 238–257.
11. Zalnezhad, E.; Baradaran, S.; Bushroa, A.R.; Sarhan, A.D. Mechanical property enhancement of Ti-6Al-4V by multilayer thin solid film Ti/TiO_2 nanotubular array coating for biomedical application. *Metall. Mater. Trans. A* **2014**,*45*, 785–797.
12. Oskouei, R.H.; Sushimatha, K. Parylene coatings in medical devices and implants: A review. *Univers. J. Biomed. Eng.***2015**, *3*, 9–14.
13. Tadic, D.; Peters, F.; Epple, M. Continuous synthesis of amorphous carbonated apatites. *Biomaterials* **2002**, *23*, 2553–2559.
14. Ahmad Ramli, R.; Adnan, R.; Abu Bakr, M.; Massudi, S. Synthesis and characterisation of pure nanoporous hydroxyapatite. *J. Phys. Sci.* **2011**, *22*, 25–37.
15. Faig-Martí, J.; Gil-Mur, F.J. Hydroxyapatite coatings in prosthetic joints. *Rev. Esp. Cir. Ortop. Traumatol. Engl. Ed.* **2008**,*52*, 113–120.
16. Kuroda, K.; Okido, M. Hydroxyapatite coating of titanium implants using hydroprocessing and evaluation of their osteoconductivity. *Bioinorg. Chem. Appl.* **2012**, *2012*.
17. Liang, H.; Shi, B.; Fairchild, A.; Cale, T. Applications of plasma coatings in artificial joints: An overview. *Vacuum* **2004**,*73*, 317–326.
18. Holmes, R.; Bucholz, R.W.; Mooney, V. Porous hydroxyapatite as a bone-graft substitute in metaphyseal defects. A histometric study. *J. Bone Jt. Surg. Br. Vol.* **1986**, *68*, 904–911.
19. Cook, S.D.; Kay, J.F.; Thomas, K.A.; Jarcho, M. Interface mechanics and histology of titanium and hydroxylapatite-coated titanium for dental implant applications. *Int. J. Oral Maxillofac. Implants* **1987**, *2*, 15–22.

20. Ozawa, S.; Kasugai, S. Evaluation of implant materials (hydroxyapatite, glass-ceramics, titanium) in rat bone marrow stromal cell culture. *Biomaterials* **1996**, *17*, 23–29.
21. Morks, M.F.; Kobayashi, A.; Fahim, N.F. Abrasive wear behavior of sprayed hydroxyapitite coatings by gas tunnel type plasma spraying. *Wear* **2007**, *262*, 204–209.
22. Metikoš-Hukovič, M.; Tkalčec, E.; Kwokal, A.; Piljac, J. An in vitro study of ti and ti-alloys coated with sol–gel derived hydroxyapatite coatings. *Surf. Coat. Technol.* **2003**, *165*, 40–50.
23. Balani, K.; Anderson, R.; Laha, T.; Andara, M.; Tercero, J.; Crumpler, E.; Agarwal, A. Plasma-sprayed carbon nanotube reinforced hydroxyapatite coatings and their interaction with human osteoblasts *in vitro*. *Biomaterials* **2007**,*28*, 618–624.
24. Apachitei, I.; Lonyuk, B.; Fratila-Apachitei, L.E.; Zhou, J.; Duszczyk, J. Fatigue response of porous coated titanium biomedical alloys. *Scr. Mater.* **2009**, *61*, 113–116.
25. Vadiraj, A.; Kamaraj, M. Fretting fatigue studies of titanium nitride-coated biomedical titanium alloys. *J. Mater. Eng. Perform.* **2006**, *15*, 553–557.
26. Lynn, A.K.; DuQuesnay, D.L. Hydroxyapatite-coated Ti-6Al-4V: Part 1: The effect of coating thickness on mechanical fatigue behaviour. *Biomaterials* **2002**, *23*, 1937–1946.
27. Lynn, A.K.; DuQuesnay, D.L. Hydroxyapatite-coated Ti-6Al-4V: Part 2: The effects of post-deposition heat treatment at low temperatures. *Biomaterials* **2002**, *23*, 1947–1953.
28. *Standard Practice for Surface Preparation and Marking of Metallic Surgical Implants*; ASTM F86-04; ASTM International: West Conshohocken, PA, USA, 2004.

Chapter 2

Influence of the Hardfacing Welds Structure on Their Wear Resistance

*Janette Brezinová *, Dagmar Draganovská, Anna Guzanová, Peter Balog and Ján Viňáš*

Department of Mechanical Technology and Materials, Technical University of Košice, Mäsiarska 74, 040 01 Košice, Slovak Republic

ABSTRACT

The contribution presents the research results of hardfacing metals' resistance in conditions of abrasive wear. Two types of hardfacing electrodes with a different chemical composition were used in the creation of three layers of hardfacing metals. The chemical composition of electrodes determines the difference in a hardface deposit structure. We have investigated the influence of mixing the base metal and a filler metal and the influence of hardfacing welds structure on the resistance against abrasive wear. The results of the experiments have showed that the intensity of wear is very dependent on the parameters of wear as well as the morphology structure of hardfacing metals.

Keywords: hardfacing welds; structure; abrasive wear; wear resistance

1. INTRODUCTION

The wear of mechanical parts of machinery is still a current scientific, engineering and economic issue. The analysis of machine parts and structure faults shows they are often caused by tribological processes that take place at the functional surfaces. The interaction between functional surfaces in their relative motion causes adverse effects in the surface layers, leading to their deterioration. An external manifestation of this process is the removal or relocation of the functional surface particles. This process can be measured by a change in the size, weight or mechanical and physico-chemical properties. This process can be alleviated by improving the structural arrangement of nodes, the appropriate choice of materials or the creation of new surface layers [1,2,3,4,5]. When creating the surface layers for tribological use, we often encounter hardfacing technology which enables not only the restoration of the worn surface geometry but gives it new, often better properties than the properties of the original material, which leads to a prolongation of their life [6,7,8].

Welding consumables can be divided into groups according to their properties and wear resistance. The iron-based alloys include martensitic alloys, austenitic alloys and alloys with a high content of carbides [9]. Martensitic alloys are used for the restoration of shape and for hardfacing. Their main advantages include good resistance to wear of metal-metal type, good impact resistance and acceptable abrasion resistance.

Austenitic claddings are suitable for the restoration of shape and are characterized by an excellent impact resistance and an acceptable abrasion resistance.

The advantage of alloys with high carbide content is both the excellent abrasion resistance, good heat resistance, and acceptable corrosion resistance; however, the disadvantage is a weak impact resistance [10].

Cobalt- and nickel-based alloys are resistant to most types of wear. Due to their high cost, they are only used in cases when their properties are economically justified, such as in high temperature applications when the iron-based alloys with a high carbide content lack sufficient resistance. Nickel alloys are a slightly cheaper alternative [11].

The most progressive additive materials for welding and thermally sprayed coatings for the conditions of abrasive, erosive, corrosive and combined stresses in mining, energy and building industries include nanostructural claddings [12]. Characteristics of these materials not containing cobalt or nickel are based on the high content of carbide-forming and boride-forming metals Cr, Mo, W and Nb, carbon (up to 5%) and boron (5%–10%). They have an amorphous metallic glass character. Very hard and very fine nanostructures with a grain size of 2–75 nm are formed during their devitrification. A hardness of 900–1230 HV, 54–74 HRC, respectively, is reached by the resulting cladding. The base element in these alloys is iron even though, in the maximum content of alloying elements, it does not reach 50%. The claddings are carried out by MIG (Metal Inert Gas) hardfacing, and coatings are carried out by HVOF (High Velocity Oxygen Fuel) and plasma spraying technology.

Essential scientific knowledge of cladding tribology is detailed in [13,14] and others. These imply that one of the determining factors for the wear intensity in both the abrasive and erosive wear of claddings is their hardness. The hardness of cladding is a function of its chemical composition, the welding heat mode during hardfacing and its structure.

In the current analysis of abrasion resistance, the issue of cladding structure and substructure and their influence on abrasion resistance is not analyzed in detail. So far there is no consensus on the most appropriate type of structure in terms of resistance to abrasive wear. Most authors coincide in finding that the resistance to abrasive wear of alloys with the same hardness with a different chemical and structural composition is not the same. This depends on the hardness, the amount, the shape, the size and the distribution of structural components. Some authors consider austenitic-carbidic the most advantageous structure, others prioritize a martensitic-carbidic structure. This results from the diversity of the abrasive wear process and a wide range of real operating conditions [6,9,15,16,17].

Structural and substructural properties can directly affect material removal. Its intensity is conditioned by the strength and cohesive properties of the submicroscopic particles of structure that are loaded by wear and abrasion. Every structural component in the process of exploitation determines the level of resistance of the whole metal with its share. In abrasive wear, two crucial stages must be distinguished. The first one is the process of forcing the abrasive into the surface where the limiting factor is the indentation hardness. The second stage is the process of surface disintegration, where the crucial role belongs to the strength of interatomic bonds and the strength of the structural components' connection to each other at the grain boundaries [13].

In real conditions, the failure of the material surface layers by a high-cycle contact fatigue process (abrasive particles do not notch into the surface, causing only elastic deformation), a low-cycle contact fatigue process (a plastic deformation when notching abrasive particles into the surface) and the grooving together with the segregation of worn material particles may occur at various locations of the tough material-worn surface. At a high speed of abrasive particles' relative motion against the worn surface, it is necessary to also consider other degradation mechanisms: the heat affection of the material (tempering of steel, softening of the polymeric material), adsorption failure (reactions of the worn material with surface active agents that reduce surface hardness) and tribochemical reactions of the worn material with the environment. Although in abrasive wear of brittle materials plastic deformation occurs, the brittle fracture often determines the rate of wear. In tough materials this fault occurs probably right after the abrasive particle, due to tensile stress, acts here. The removal of the material by abrasion in brittle materials occurs by a brittle fracture rather than by the plastic deformation. Also, in the abrasive wear of heterogeneous materials which contain tough and brittle phases, infringement by brittle fracture may occur. The predominant mechanism of material removal depends on the characteristics of individual phases and their volume fractions [14].

The stated information implies that due to the complexity of the wear process, it is necessary to design a type of cladding based on the tribological analysis and the conditions of the surface stress. This analysis, however, is for many practical cases impracticable due to the imperfection of developed procedures and, therefore, operational and laboratory tests have recently been used for the evaluation of cladding properties.

The aim of the paper is to carry out a tribological and metallographical analysis for two types of hardfacing layers. The analysis can contribute to clarification of the relationship between the microstructure and wear resistance of layers.

2. MATERIALS AND METHODS

The base material for production of hardfacing samples also used as a gauge in the tests of wear was steel of grade S235JRH. Dimensions of the test samples were 20 × 20 × 8 mm. This is a structural carbon steel with 0.22% C and excellent weldability.

For cladding formation hardfacing electrodes E 508 B and E 518 B (Welding Research Institute, Industrial Institute of Slovakia, Bratislava, Slovak Republic) were used, whose chemical composition meets the requirements for heterogeneity of structural constitution of claddings. Chemical composition of electrodes is shown in Table 1. Technological and use properties are as follows:

- E 508 B—thick flux-covered electrode. Recommended preheating: without preheating or preheating up to 250 °C, depending on the quality and dimensions of the hardfacing component. It is used for hardfacing of working parts of agricultural and forestry machines exposed to abrasion. The cladding is resistant to mild impacts, it is not thermally treated. If any treatment is needed, it shall be spheroidizing annealed and subsequently thermally treated to optimal hardness of approximately 600 HV (Vickers hardness).
- E 518 B—thick flux-covered electrode. Cladding preferably alloyed from the flux cover. Preheating for hardfacing is 400 °C with following cooling in wrap or in the oven. Cladding is not thermally treated. It is used for the parts of mining and earth-moving machinery, metallurgical equipment exposed to abrasive and erosive wear combined with impacts. Hardness of the cladding is 600–660 HV.

Table 1. Chemical composition and hardness of welding electrodes in wt. %.

Electrode	C	Mn	Si	Cr	Mo	HV
E 508 B	0.5	0.7	0.5	6.0	0.6	580
E 518 B	3.4	0.5	0.8	29	-	660

Cladding samples for the studies of structural condition and the fracture surfaces were made from one to three layers. After cooling the cladding was broken off by three-point bending. The surfaces obtained this way were used to study the failure mechanism and for the structural analysis. Abrasion resistance was determined with an assistance of a test device based on the principle of relative movement of test samples submerged into free abrasive [16]. For this test abrasive based on Al_2O_3 with grain size of 1.2 mm was used. Incidence angle between abrasive and test sample was 75° and speed of relative sample motion was 2.25 $m \cdot s^{-1}$. The weight losses were investigated using sensitive digital scale weight with a measurement accuracy of 10^{-4} g.

To determine the microhardness of individual structural phases, the microhardness HV0.05 was measured, with a load of 0.5 N and dwell time 10 s. The measurements were carried out using Hannemann head (Carl Zeiss AG, Feldbach, Switzerland). The course of the hardness in the cross-section of claddings was realized by the Vickers hardness pyramid (HPO 250, Liepzig, Germany) with a load of 294.3 N and dwell time 10 s.

Metallographic study and a study of fracture surfaces were realized using light microscope Olympus BXFM (Olympus Deutschland GmbH, Hamburg, Germany) and a scanning electron microscope TESLA-BS-301 (Tesla, Brno, Czech Republic).

3. RESULTS AND DISCUSSION

The course of the hardness of claddings E 508 B showed a strong effect of mixing the weld and base metal, as shown in Figure 1. The hardness of the cladding in its first layer in the area of melting rises sharply to a value of 550 HV30. The second layer reaches an average value of 615 HV30, while in three-layer cladding the hardness reaches around 630 units. Theoretical and practical assumptions that the effect of mixing the base material and weld metal does not disappear until the third layer were confirmed [13]. The transition of the hardness curves from the minimal values of the base material to the maximum values in the cladding showed a step-change.

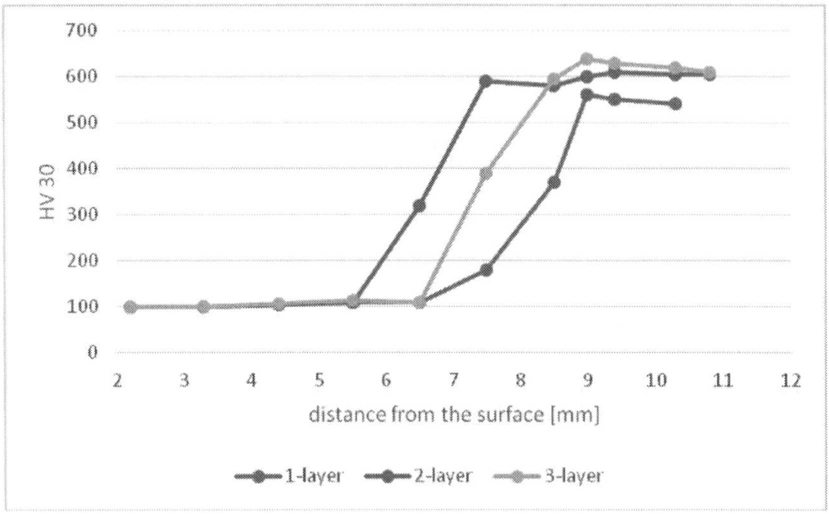

Figure 1. Course of hardness in cladding E 508 B with various numbers of layers.

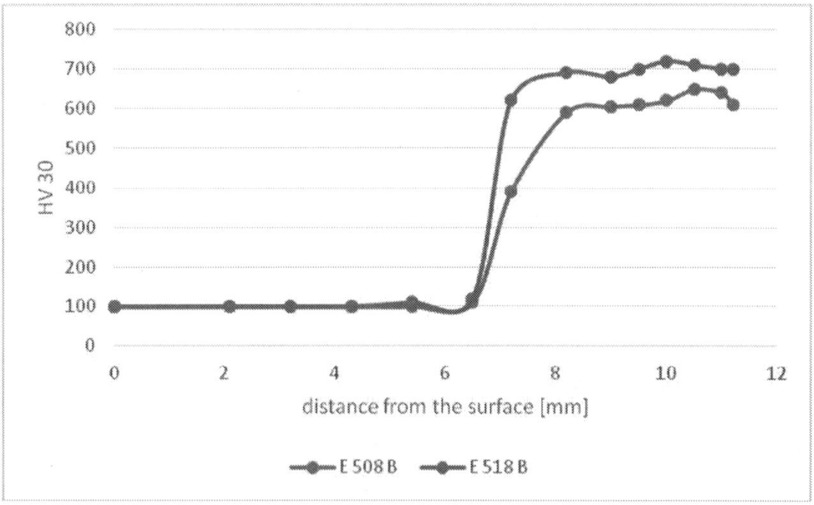

Figure 2. Course of hardness in three-layer hard-faced materials.

A similar course of the hardness was shown by cladding E 518 B, except that the hardness value stated by the manufacturer, which is 660 HV30, was reached already in the first layer. Maximum values obtained in individual layers did not differ significantly and reached up to 730 HV. Due to the high content of C and Cr, the mixing with the base material was significantly less.

The course of three-layer cladding hardness is shown in Figure 2. In terms of achieved hardness, higher wear resistance can be assumed in claddings realized by the E 518 B electrode.

The results of abrasion resistance tests of evaluated hardfaced materials are graphically presented in Figure 3 and Figure 4. From these figures it can be seen that, in electrode E 508 B, single-layer claddings show higher mass loss due to the mixing of the weld metal with low-carbon steel. Despite the fact that for electrode E 518 B small differences of hardness values were reached in individual layers, there is a big difference in mass loss between single-layer and multiple-layer claddings. Compared to electrode E 508 B, it reaches higher values. The results of mass loss values of tested claddings cannot be evaluated just by a chemical composition and measured values of hardness. Most authors explain the theory of material removal in wear by a contact interaction where a plastic deformation occurs [6]. This deformation is caused by the redistribution and dissipation of energy. The energy redistribution, followed by implementation of plastic deformation, causes a movement of anchored and free dislocations. The dislocation state of metals depends mainly on the structural state. The real structural state is therefore another factor influencing the wear resistance [13,15].

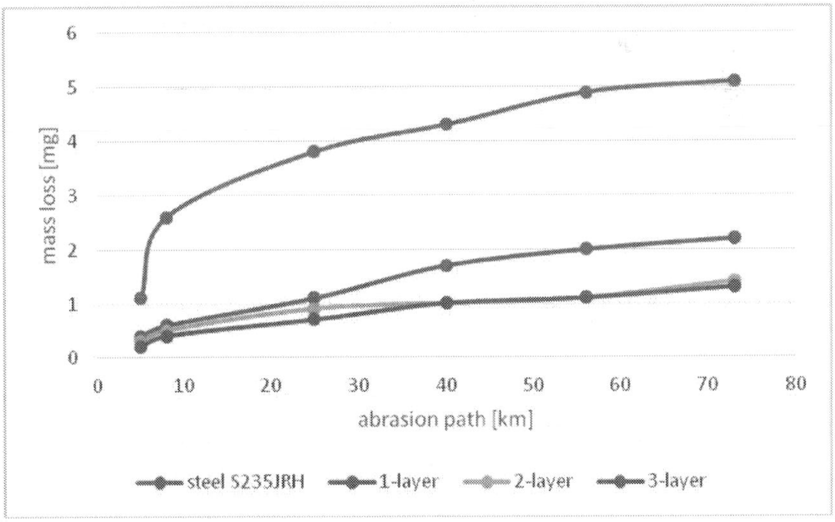

Figure 3. Course of wear for hardfacing material E 508 B.

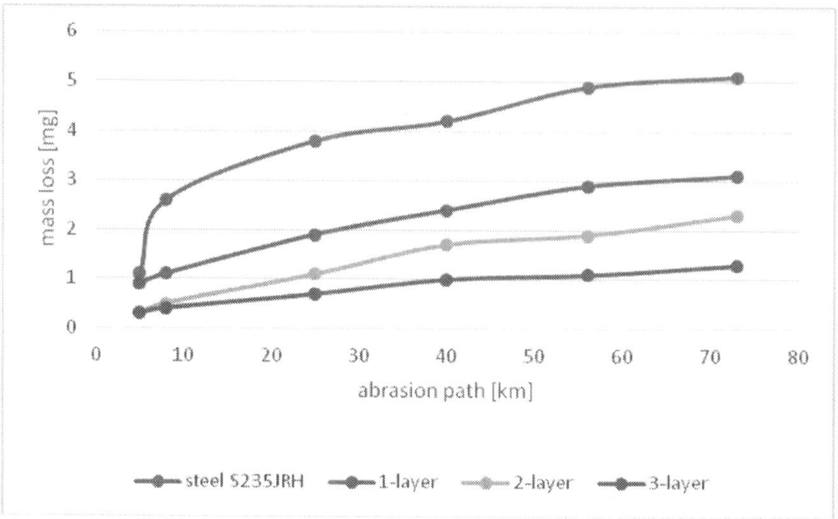

Figure 4. Course of wear for hardfacing material E 518 B.

As mentioned above, the investigated claddings vary significantly in internal structure. Electrode E 508 B represents cladding based on mild steel with a content of C up to 0.4%. The relatively high content of carbide-forming elements Cr and Mo determines the formation of the bainitic matrix structure and, by bonding with carbon, creates disperse carbides reinforcing the very fine bainitic matrix. Since claddings are multilayered and multipass in a layer, the mutual thermal influence causes very heterogeneous formation of the phases. In Figure 5 there is a structure of the first layer in the area of the mixing of the weld metal with the base material which was not thermally affected. It is a very fine bainitic structure with a ferrite net which retains the oriented nature. The microhardness of this layer reaches 424 HV 0.05. In the area of thermal affection, the grains are finer and the distinctive casting character is disappearing, as shown in Figure 6. The similar nature of the structure is also present in the second layer, where the hardness reaches 489 HV 0.05. In the third layer of the cladding, which was not thermally affected, the distinctive casting structure is retained, as shown in Figure 7 (538 HV 0.05). The cladding structure in the third layer in the thermally affected zone is shown in Figure 8. In the bainitic matrix, the very dispersely formed carbidic phase can be observed. Claddings are thus characterized by a relatively tough matrix in which the very disperse carbidic phase ensures sufficiently high hardness and, therefore, abrasion resistance.

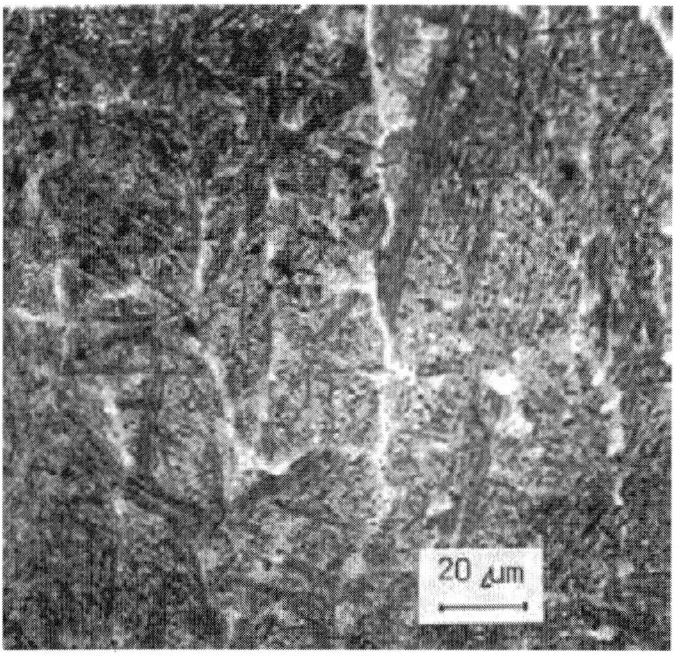

Figure 5. Microstructure of hardfacing weld E 508 B.

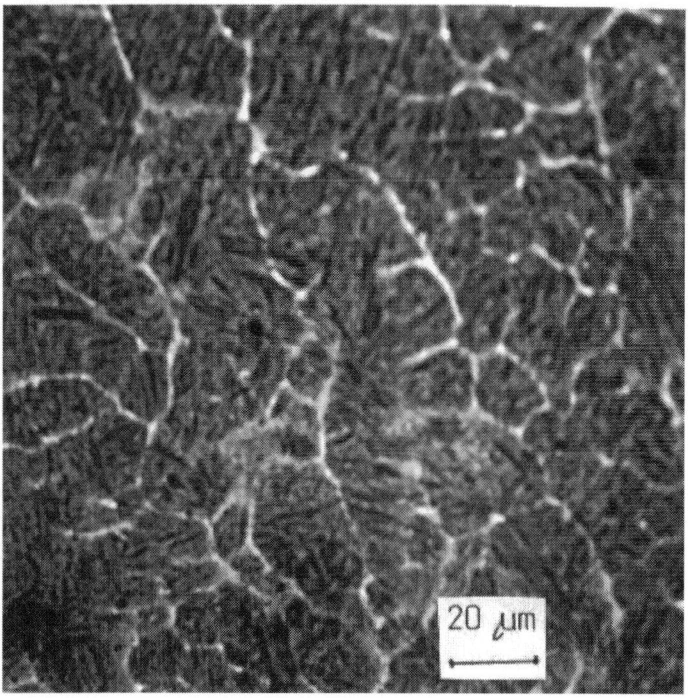

Figure 6. Thermally affected structure of the cladding E 508 B.

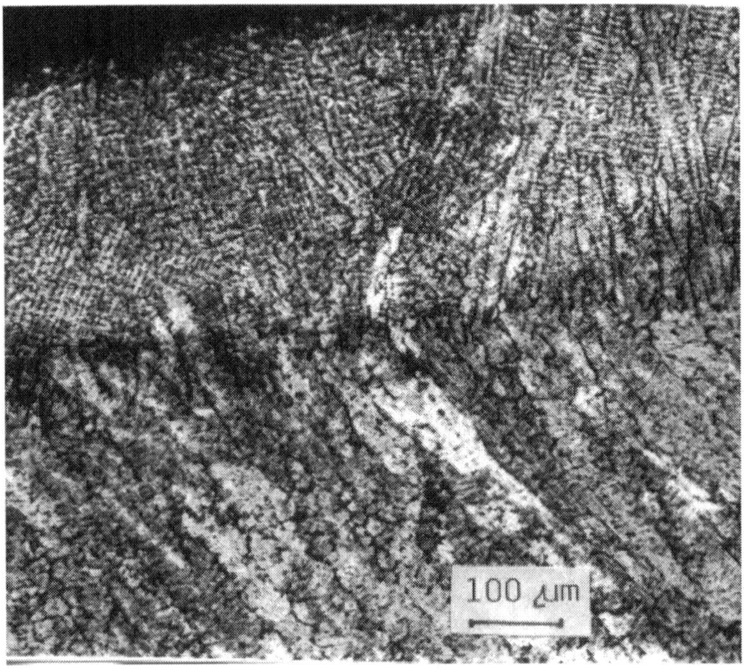

Figure 7. Cast structure of hardfacing weld E 508 B.

The structural composition of the claddings realized by electrode E 518 B is shifted into the area of cast iron, which is manifested by the significant dendritic character of the cast structure. The high content of C up to 3.5% and chrome up to 27.5% after the re-melting in the cladding gives a high-alloyed cast iron of ledeburitic type with a significant presence of primary and secondary carbides. The structure in the transition from the base material to the cladding is documented in Figure 9. Figure 9 shows that even in the short term of the re-melting, a very thin carburized layer was formed in the base material, demonstrated by large pearlitic grains. In the areas of the cladding where the cast structure was thermally affected, a very fine, oriented needle structure (Figure 10) or a transformed structure with large carbidic needles (Figure 11—TMP14) was formed due to the structural transformation. The structure in the cover layers of the cladding is a casting structure with a distinct dendritic composition (Figure 12). The average hardness of the ledeburitic matrix was of 562 HV 0.05. Large carbidic needles reached a hardness of 1630 HV 0.05 while small needles reached up to 2350 HV 0.05.

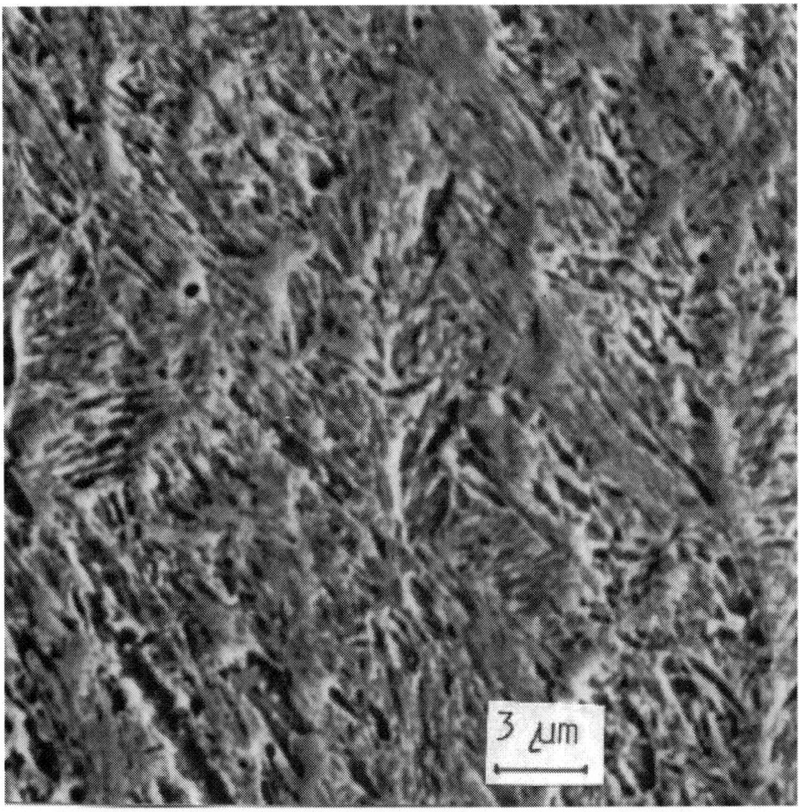

Figure 8. Carbidic phase in bainitic matrix of the weld E 508 B.

For the creation of the fracture surface, energy is demanded and, thus, the character of the fracture, whether it is cleaving, ductile failure or decohesion along the phase boundary, also provides the resistance to separation of microparticles under surface abrasion. Therefore, the examination of these surfaces makes it possible to predict the use properties of metallic materials.

Depending on the heat affection of the structural composition, the bulk of the fracture surfaces of E 508 B under bending stress was formed by decohesive failure at the border of crystallites or dendrites, as shown in Figure 13. Part of the surface failed by decohesion with a low power consumption—smooth facets and part of the surface failed in these claddings by pit morphology whose creation is energy-demanding and indicates the certain resistance to a brittle fracture, as shown in Figure 14.

Figure 9. Cast structure of hardfacing weld E 518 B.

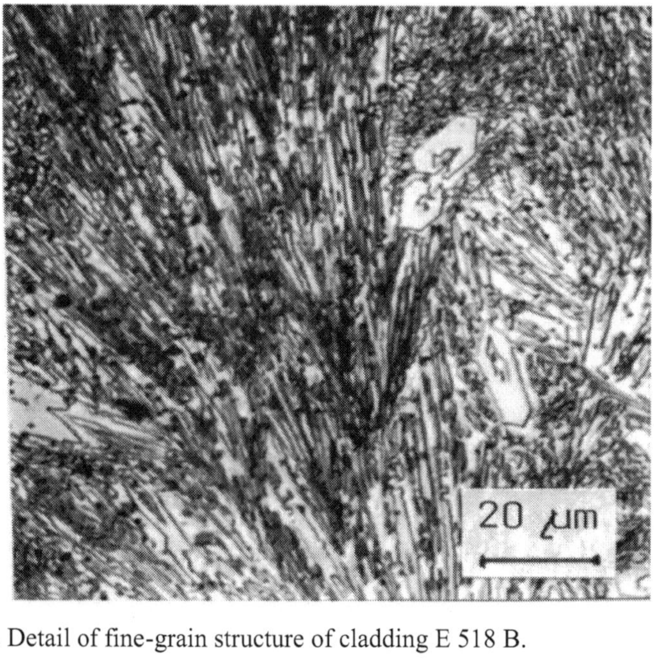

Figure 10. Detail of fine-grain structure of cladding E 518 B.

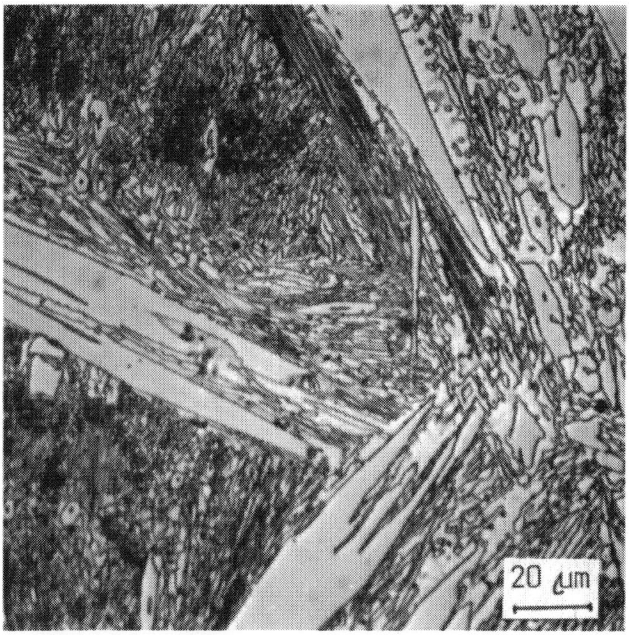

Figure 11. Detail of massive carbidic particles E 518 B.

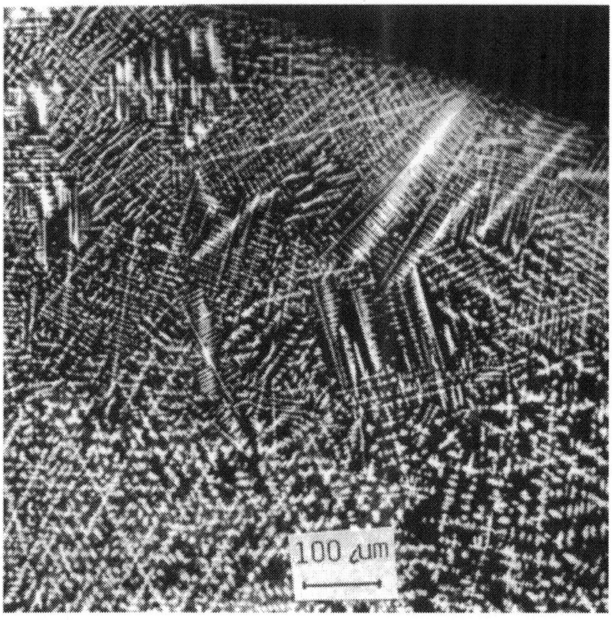

Figure 12. Cast structure of hardfacing weld E 518 B.

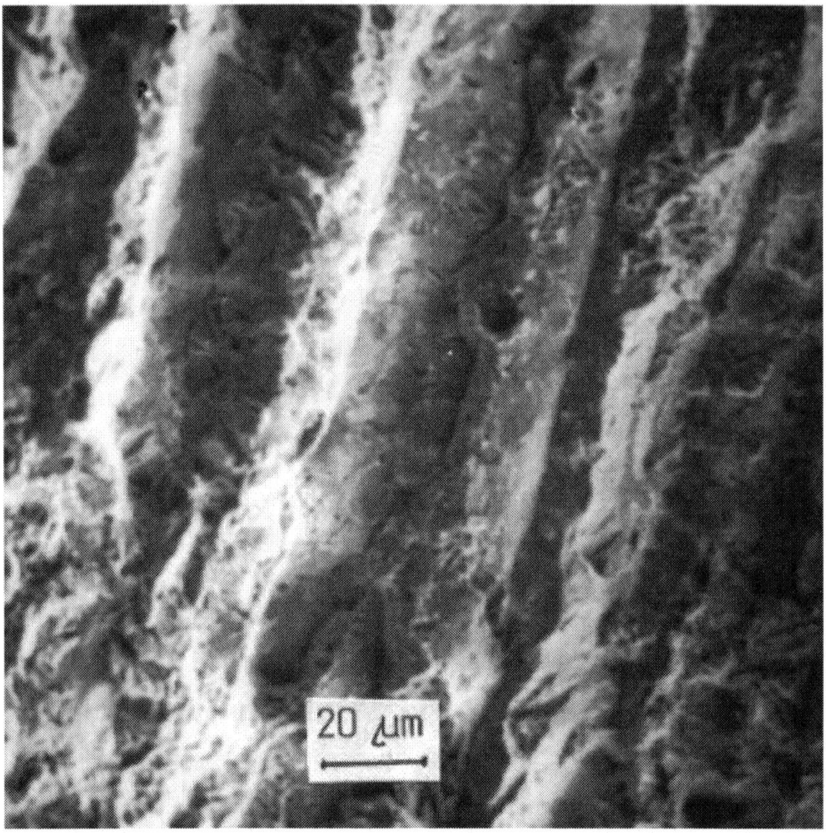

Figure 13. Decohesive failure at the dendrite borders.

The fracture surfaces of claddings of E 518 B obtained for the purpose of studying the mechanism of their formation confirm that the energy demand for their creation is very low. The view of the fracture surface is shown in Figure 15. Part of the fracture surface consists of decohesive facets without morphological signs of cleaving or ductile failure. Their formation has occurred by the simple separation of the surfaces at the grain borders with minimum energy consumption. A large part of the fracture, however, consists of facets of cleavage decohesion. These are facets formed by the separation of phase boundaries, but the mechanism of their failure is the cleavage, *i.e.*, low energy. These surfaces are shown in Figure 16 and are characteristic for the cast state of the cladding without thermal treatment.

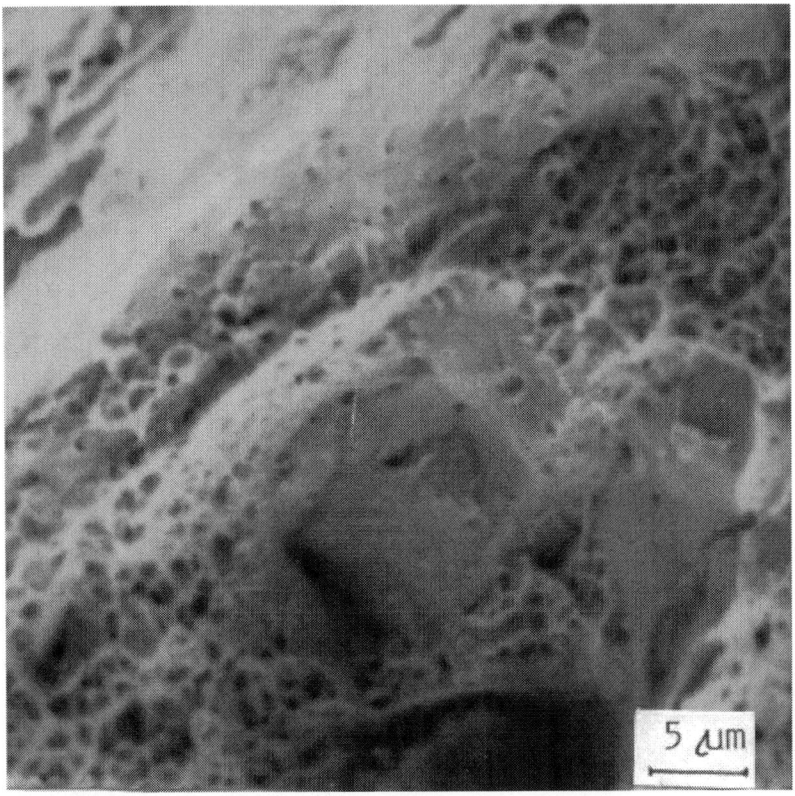

Figure 14. Morphology of E 508 B cladding failure.

The stated information implies that claddings made by electrode E 508 B have a tough bainitic matrix dispersely reinforced by a carbidic phase, as shown in Figure 17. This symbiosis of the matrix and disperse carbidic reinforcement optimizes the abrasive wear resistance. The structure has a high density of dislocations whose movement is intensively stopped by the disperse phase, as shown in Figure 18. Plastic deformation thus requires high stress for the movement of the dislocations. It is known that the bainitic structures with the same chemical compositions optimize mechanical properties,*i.e.*, at high levels of strength (hardness) they retain adequate ductility.

The structure of claddings made by electrode E 518 B is ledeburitic due to the high content of carbon and the high content of chrome. There are primary as well as secondary carbides present in the structure, morphologically formed massively as well as dispersely, a shown in Figure 19. The dislocation state is high, represented mainly by anchored dislocations.

Figure 15. Fracture surface of cladding E 518 B.

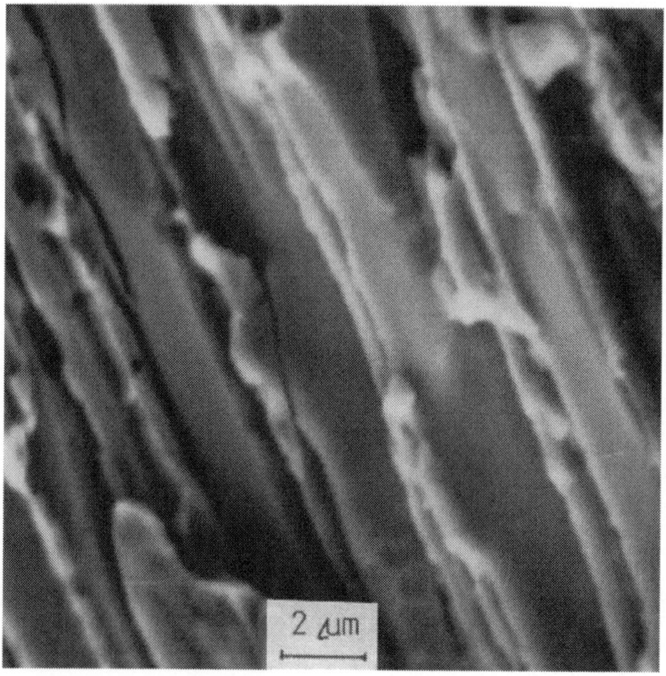

Figure 16. Detail of decohesive failure.

Figure 17. Bainitic matrix containing carbides.

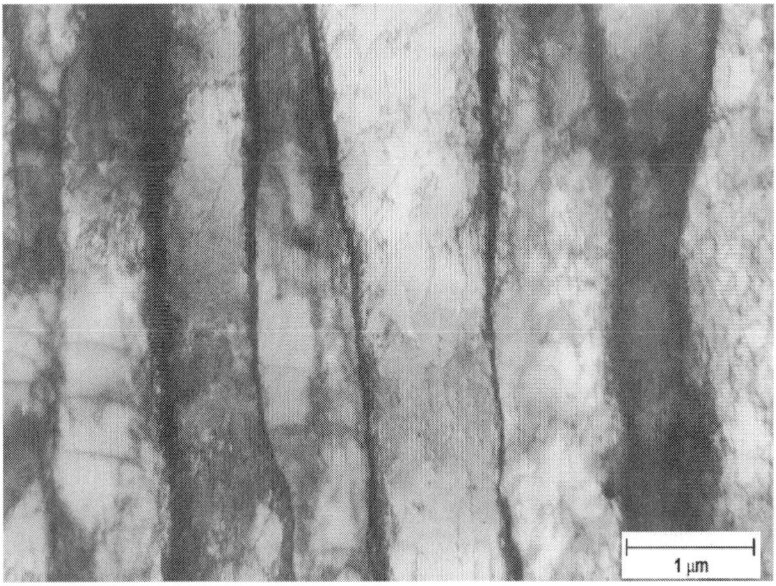

Figure 18. Dislocations in bainitic matrix.

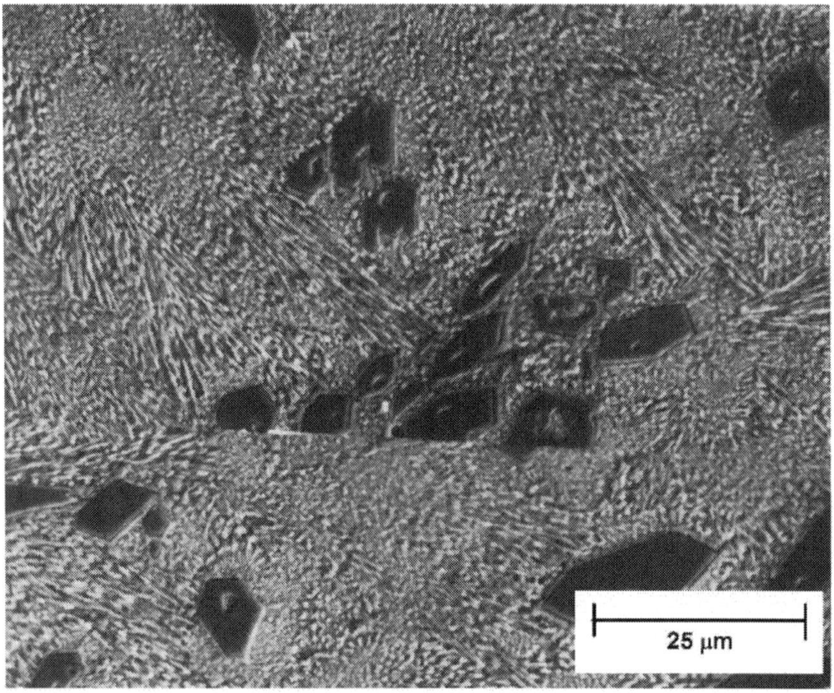

Figure 19. Carbides in ledeburitic structure.

Although the material is high strength (hard), it has a small reserve of plastic deformation, and therefore it is brittle. The low strength of the grain boundaries is also added to these properties. It can therefore be concluded that the cohesive strength of the boundaries and subboundaries is significantly lower than the strength of the matrix. The material separation is realized by the decohesion of particles and the separation by the cleaving mechanism.

4. CONCLUSIONS

Based on the achieved results of the study of the structural composition of claddings, the measurements of their hardness and their abrasive wear resistance, the following can be concluded:

1. The hardness and wear values confirm the correctness of the principle that the optimal properties are not achieved by hardfaced materials until the third layer. For practical application it is necessary to execute the welding process with minimal melting of the base material. In the cladding by

electrode E 518 B, due to its chemical composition, this effect was less pronounced.

2. In set conditions of wear, the claddings made by electrode E 508 B seem more favorable. The tough bainitic matrix dispersely reinforced by a carbidic phase provides good resistance to abrasive wear. In terms of chemical composition and manufacturing technology, these claddings appear economical since they can be realized without a special regime of hardfacing. Cladding has excellent properties without a heat treatment.
3. Claddings made by electrode E 518 B, in terms of hardness values, appear to be of high quality. The formation of the ledeburitic structure with plenty of carbidic phase gives a high-strength though brittle cladding. Claddings therefore have little resistance to abrasive wear in combination with mild impacts.
4. Based on the obtained results, it can be concluded that in the optimized chemical composition of weld materials, it is necessary to also take the structural constitution of the cladding into consideration. Better results are obtained in two-component structural phases where the matrix has sufficient hardness with a good supply of plastic properties and the reinforcing disperse component is based on a carbidic or carbonitridic basis.

ACKNOWLEDGMENTS

This work was supported by the Slovak Research and Development Agency under the contract No. SK-UA-2013-0013 and project VEGA No. 1/0600/13.

AUTHOR CONTRIBUTIONS

Realization of hardfacing, three-point bending, abrasive tests—Ján Viňáš, Peter Balog. Metallography analysis, analysis of fracture, hardness testing—Janette Brezinová, Dagmar Draganovská, Anna Guzanová.

REFERENCES

1. Viňáš, J.; Brezinová, J.; Guzanová, A.; Svetlík, J. Degradation of renovation layers deposited on continuous steel casting rollers by submerged arc welding. *Proc. Inst. Mech. Eng. B* **2013**, *227*, 1841–1848.
2. Viňáš, J.; Brezinová, J.; Guzanová, A. Analysis of the quality renovated continuous steel casting roller. *Sadhana* **2013**,*38*, 477–490.
3. Viňáš, J.; Brezinová, J.; Guzanová, A.; Balog, P. Evaluation of the quality of cladding deposited on continuous steel casting rolls. *Int. J. Mater. Res.* **2013**, *104*, 183–191.

4. Correaa, E.O.; Alcântara, N.G.; Valeriano, L.C.; Barbedo, N.D.; Chaves, R.R. The effect of microstructure on abrasive wear of a Fe-Cr-C-Nb hardfacing alloy deposited by the open arc welding process. *Surf. Coat. Technol.* **2015**, *276*, 479–484.
5. Jankauskas, V.; Antonov, M.; Varnauskas, V.; Skirkus, R.; Goljandin, D. Effect of WC grain size and content on low stress abrasive wear of manual arc welded hardfacings with low-carbon or stainless steel matrix. *Wear* **2015**, *328–329*, 378–390.
6. Blaškovič, P.; Čomaj, M. *Renovácia Naváraním a Žiarovým Striekaním*; Alfa: Bratislava, Slovakia, 1991.
7. Dwivedi, D.K. Abrasive wear behavious of iron based hard serfacing alloy coatings developed by welding. *Surf. Eng.***2004**, *20*, 87–92.
8. Senthilkumar, B.; Kannan, T. Effect of flux cored arc welding process parameters on bead geometry in super duplex stainless steel claddings. *Measurement* **2015**, *62*, 127–136.
9. Czichos, M. *Tribology*; Elsevier: Oxford, UK; New York, NY, USA, 1978.
10. Marinescu, I.D.; Hitchiner, M.P.; Uhlmann, E.; Rowe, W.B.; Inasaki, I. *Handbook of Machining with Grinding Wheels*; CRC Press: New York, NY, USA, 2007; p. 593.
11. Malkin, S.; Guo, C. *Grinding Technology: Theory and Applications of Machining with Abrasives*; Industrial Press: New York, NY, USA, 2008; p. 372.
12. Abušinov, A. Nanostrukturní návary a žárové nástřiky. *MM Průmyslové Spektrum* **2011**, *4*, 22–24.
13. Brožek, M. Výsledky zkoušek abrazivního opotřebení vrstev. In Proceedings of the Conference Renop 92, Trnava, Slovakia, 17–19 March 1992; pp. 139–142.
14. Hawk, J.A.; Wilson, R.D. *Abrasive Wear Failures*; ASM Handbook; ASM International: Materials Park, OH, USA, 2002; Volume 11, pp. 906–921.
15. Adamka, J.; Petríková, G. Vplyv štruktúry návarov na odolnosť proti abrazívnemu opotrebeniu. In Proceedings of the Conference Intertribo 93, Bratislava, Slovakia, 26–28 August 1993; pp. 70–76.
16. Jankura, D. Tribologické vlastnosti viacvrstvových tvrdonávarov. In Proceedings of the Conference Funkčné Povrchy, Trenčín, Slovakia, 14–16 June 2001; pp. 96–102.
17. Jankura, D. Hodnotenie vlastností viacvrstvových tvrdonávarov. *Mechanika* **1996**, *47*, 59–66.

Chapter 3

Enhancing Corrosion and Wear Resistance of AA6061 by Friction Stir Processing with $Fe_{78}Si_9B_{13}$ Glass Particles

Lingyu Guo[1], *Yan Liu*[1], *Kechang Shen*[1], *Chaoqun Song*[1], *Min Yang*[1], *Kibuem Kim*[2] *and Weimin Wang*[1,*]

[1] *Key Laboratory for Liquid-Solid Structural Evolution and Processing of Materials, Ministry of Education, Shandong University, Jinan 250061, China*
[2] *Hybrid Materials Center (HMC), Faculty of Nanotechnology and Advanced Materials Engineering, Sejong University, 209 Neungdong-ro, Gwangjin-gu, Seoul 143-747, Korea*

ABSTRACT

The AA6061-T6 aluminum alloy samples including annealed $Fe_{78}Si_9B_{13}$ particles were prepared by friction stir processing (FSP) and investigated by various techniques. The $Fe_{78}Si_9B_{13}$-reinforced particles are uniformly dispersed in the aluminum alloy matrix. The XRD results indicated that the lattice parameter of α-Al increases and the preferred orientation factors *F* of (200) plane of α-Al reduces after friction stir processing. The coefficient of thermal expansion (CTE) for FSP samples increases at first with the temperature but then decreases as the temperature further increased, which can be explained by the dissolving of Mg and Si from β phase and $Fe_{78}Si_9B_{13}$ particles. The corrosion and wear resistance of FSP samples have been improved compared with that of base metal, which can be attributed to the reduction of grain size and the CTE mismatch between the base metal and reinforced particles by FSP, and the lubrication effect of $Fe_{78}Si_9B_{13}$ particles also plays a role in improving wear resistance. In particular, the FSP sample with reinforced particles in amorphous state exhibited superior corrosion and wear resistance due to the unique metastable structure.

Keywords: aluminum alloy; $Fe_{78}Si_9B_{13}$; friction stir processing; corrosion; wear

1. INTRODUCTION

Aluminum and its alloys are used extensively in numerous fields due to their low densities and high strength to weight ratio [1,2]. Nevertheless, their poor resistance to wear and corrosion causes certain limitations for their application [3]. Metal matrix composites are a new type of material that exhibit good wear and corrosion resistance properties as compared to the matrix [4,5].

As wear and corrosion are surface-dependent degradation, they can be improved by a suitable modification of surface microstructure and/or composition. A proper technique can be employed to refine the microstructure and homogeneous dispersion of reinforcements on metallic surface. With conventional surface modification techniques it is difficult to improve the dispersion of reinforcement particles on the metal surface [6]. According to previous research [7,8], thermal spraying and laser beam techniques were used to prepare surface composites, but the creation of an unfavorable phase contributed to the degeneration of the properties. These techniques were operated at high temperature and it was difficult to avoid a reaction between the reinforcements and the matrix, which leads to the formation of an adverse phase. Thus, a fabrication process below the melting temperature of the matrix is critical to avoid the complications mentioned above.

Recently, friction stir processing (FSP), a successful surfaced modification technique derived based on the principles of friction stir welding (FSW) by Mishra *et al.* [9], has attracted much attention [10,11]. During the processing, a non-consumable rotating tool is plunged into the interface between two plates to be joined and traversed along the line of the joint with a specially designed pin and shoulder. The severe plastic deformation and material flow in the stirred zone can be utilized to achieve bulk alloy modification via mixing of other elements or the second phase into the stirred alloys, accompanied with significant grain refinement [9]. Friction stir processing is a solid state processing technique to obtain surface composites. In previous studies [12,13,14,15], the most widely used reinforcements are ceramics, such as Al_2O_3 or SiC in the conventional metal matrix composites. There is little research about using amorphous alloy as a reinforced material. The amorphous materials have superior properties such as high strength, high hardness, and excellent corrosion resistance due to their disordered and metastable structure [16,17,18,19].

In the present study, we focus on the development of a new AA6061-T6 alloy matrix reinforced with $Fe_{78}Si_9B_{13}$ particles, which are fabricated by a friction stir process. The microstructure, corrosion resistance, wear behavior, and magnetic properties of the composites and base metal are investigated by different techniques.

2. EXPERIMENTAL PROCEDURE

In this study, the base metal employed is a 4 mm thick aluminum alloy AA6061-T6, which has a chemical composition of (in wt.%) Mg: 0.95; Si: 0.54; Fe: 0.22; Mn: 0.13; Cu: 0.17; Cr: 0.09; Zn: 0.08; Ni: 0.02; Ti: 0.01; Al: balance [20]. Using manual drilling machines we created a series of blind holes in the aluminum alloy surface, with a diameter and depth of 3 mm. The reinforced particles are $Fe_{78}Si_9B_{13}$ metal glasses after annealing and high energy milling processes. The annealing temperatures are 300 °C and 700 °C, respectively. The milling parameters are selected as following: the ratio of ball to powder is 10:1, frequency is 20 Hz, and the milling temperature is −196 °C. The FSP machine was used for fabricating *in situ* hybrid surface composites. The parameters for FSP include a traveling speed (v) of 45 mm/min, a rotational rate (w) of 900 r/min, a descending distance of 3 mm, and four stirrings. The base metal and the composites reinforced with particles annealed at 300 °C and 700 °C were named S000, S300, and S700, respectively.

The phase components of samples were investigated by X-ray diffractometer with Cu target (D/MAX 2500/PC, Rigaku, Tokyo, Japan, Cu Kα, λ = 0.154056 nm) and Co target (X'Pert PRO, PANalytical, Almelo, the Netherlands, Co Kα, λ = 0.178897 nm). The lattice constant is calculated by the extended Bragg equation [21]:

$$a_0 = \frac{\lambda}{2\sin\theta}\sqrt{h^2+k^2+l^2}\,, \tag{1}$$

where λ is the radiation wavelength; θ is the diffraction angle; and h, k,and l are the crystal plane indices. In present samples, the preferred orientation factors F of (200) plane of α-Al are investigated by the Lotgering method [22]:

$$F = \frac{P-P_0}{1-P_0},\ P = \frac{\sum I_{(h00)}}{\sum I_{(hkl)}}\ \text{and}\ P_0 = \frac{\sum I_{0(h00)}}{\sum I_{0(hkl)}}, \tag{2}$$

where $I_{(h00)}$ and $I_{(hkl)}$ are the integral intensities of diffraction peaks for composites samples and randomly oriented samples, respectively;

and P and P_0 are the ratios of integral intensities of (h00) planes to those of all (hkl) planes for composites samples and randomly oriented samples, respectively.

The microstructure of samples was examined by scanning electron microscopy (SEM, SUPRA55, Zeiss, Germany). Thermal expansion curves were measured by a dilatometer (DIL, Netzsch 402C, Netzsch, Shanghai, China) with a constant heating rate 20 K/min, performed under flowing high purity argon to protect the samples from oxidation.

The electrochemical measurements were carried out by CHI660E advanced electrochemical workstation with a scanning rate of 1 mV/s at room temperature. There was a typical three-electrode system in electrochemical measurements: working electrode, platinum counter electrode and $Hg|Hg_2Cl_2$ (SCE) reference electrode. The electrolyte is 3.5 wt% NaCl solutions.

The wear behavior of the base metal and hybrid surface composites was measured using a wear apparatus (M-2000, Yinuo Testing Machine, Jinan, China) at room temperature. The wear test was conducted at a rotation rate of abrasive wheel of 200 r/min, normal force of 120 N, and sliding time of 10 min.

3. RESULTS

Figure 1a illustrates an overview of the side of the FSP plate in contact with the shoulder tool. This surface shows semicircular features, similar to those induced by a conventional milling process without cracking and other disfigure. The XRD patterns of the reinforced particles are also shown in Figure 1b, explaining that the particles annealed at 300 °C are in the amorphous state, and those annealed at 700 °C are in the crystalline state. Figure 1c,d show the morphology of the samples with the reinforced particles annealing at 300 °C and 700 °C, denoted as S300 and S700, respectively. It can be found that the reinforced particle is relatively uniform dispersed in the aluminum alloy matrix, and the size of the particle annealed at 700 °C is smaller than that of the particle annealed at 300 °C.

Figure 2 shows the XRD patterns of AA6061 FSP samples with $Fe_{78}Si_9B_{13}$ particles (S300 and S700) as well as the base alloy (denoted S000). The main diffraction peaks are α-Al phases, and the other intermetallic phases are not found due to the less content. The (111), (200), (220), and (311) peaks of α-Al are easily identified, and the intensity of the (200) peak changed obviously. The inset shows that the peak of diffraction (2θ) of (200) plane of α-Al in S300 shifts to the left after FSP. For clarity, the angle of diffraction (2θ), lattice parameter a_0, and preferred orientation factors F of (200) plane of α-Al are listed in Table 1. According to Equation (1), the lattice constants of the (200) plane of α-Al for the three samples are in the order S300 > S700 > S000. However, the preferred orientation factors F of the (200) plane of α-Al for the samples are in the order S300 < S700 < S000, implying that the initial rolling texture was destroyed during friction stir processing. In addition, the variation of XRD data with the Co target is confirmed by that with the Cu target.

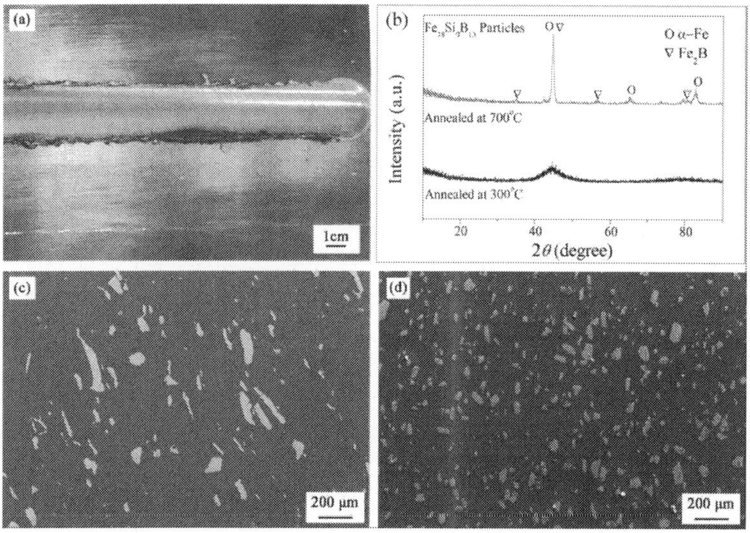

Figure 1. Topography of FSP samples: (**a**) Macrograph of samples after friction stir processing; (**b**) XRD patterns of the reinforced particles; (**c**) SEM photomicrograph of S300; (**d**) SEM photomicrograph of S700.

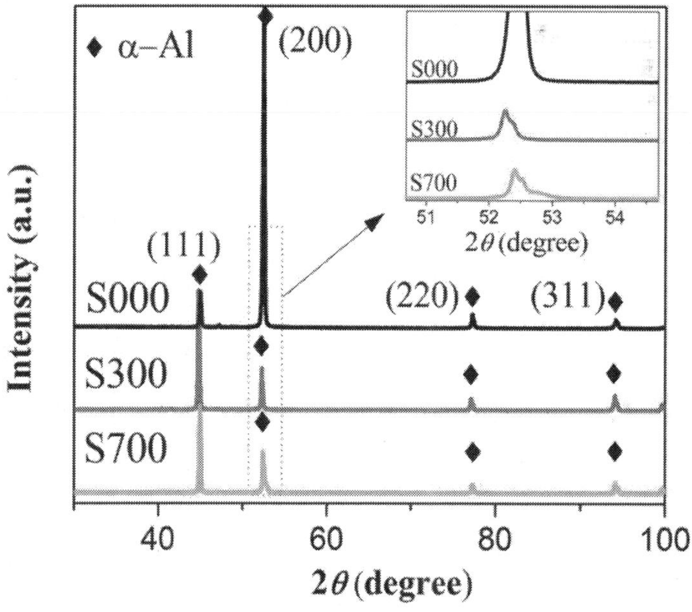

Figure 2. The XRD diffraction patterns for base metal (S000) and FSP samples (S300 and S700).

Table 1. The preferred orientation factors F, lattice parameter a_0, and angle of diffraction (2θ) of the (200) plane of α-Al of samples obtained from X-ray diffraction.

Sample	2θ (degree)	a_0 (angstrom)	$(a_0 - a_0^*)/a_0^*$ (%)	F	F'
S000	52.434	4.0495	0.003	0.7530	0.7462
S300	52.252	4.0626	0.330	0.0031	0.0293
S700	52.406	4.0515	0.052	0.1023	0.1009

Notes: 2θ, a_0, and F were tested by the Co target; F' was tested by the Cu target; and a_0^* was the lattice constant of pure Al.

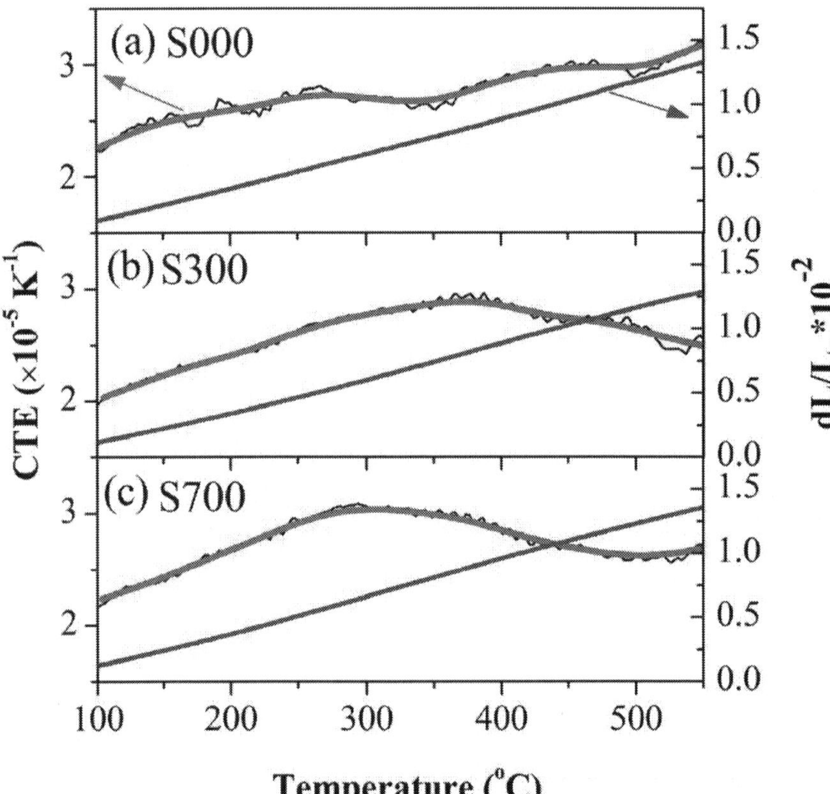

Figure 3. Thermal expansion curves (blue) and the variation of experimental coefficient of thermal expansions (CTE) (red) for (**a**) base metal (S000) as well as (**b**) S300 and (**c**) S700 FSP samples.

The thermal expansion curves and the variation of experimental coefficient of thermal expansions (CTE) obtained for base metal and FSP samples are shown in Figure 3. The three samples have an approximately linearly varying trend: the thermal expansion ratio increases with increasing temperature. In addition, the CTE of S000 increases with increasing temperature, but there is a plateau at about 300 °C. By contrast, the CTE of FSP samples at first increases with the temperature but then decreases as the temperature is further increased, and the temperature at which the CTE begins to decrease for S300 is about 350 °C, which is higher than that for S700, which is about 300 °C. Furthermore, the maximal CTEs for S300 and S700 are $2.87 \times 10^{-5}\ K^{-1}$ and $3.04 \times 10^{-5}\ K^{-1}$, respectively, and that of S000 at platform is $2.71 \times 10^{-5}\ K^{-1}$, which is lower than the former values. Moreover, at 550 °C, the CTE of S000 is about $3.18 \times 10^{-5}\ K^{-1}$, which is higher than that of S300 and S700, which are $2.48 \times 10^{-5}\ K^{-1}$ and $2.71 \times 10^{-5}\ K^{-1}$, respectively, *i.e.*, in the order S300 < S700 < S000.

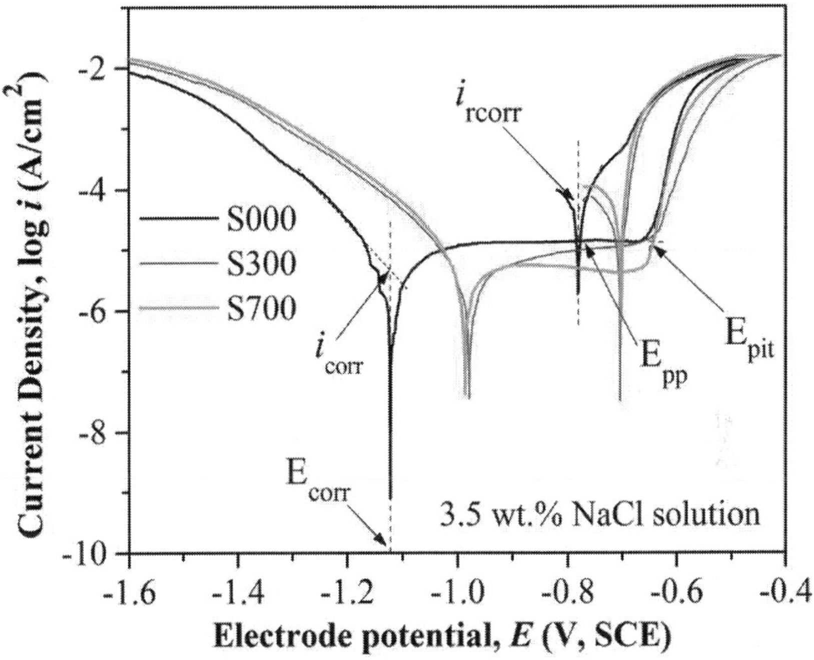

Figure 4. Cyclic anodic polarization curves of base metal (S000) and FSP samples (S300 and S700).

The cyclic anodic polarization curves of the base metal and FSP samples with a scanning rate of 1 mV/s, obtained in 3.5 wt% NaCl solutions, are shown in Figure 4. As can be seen from the curves, an obvious current plateau is observed in each curve, which is associated with passive film formation during the anodic polarization. The electrochemical parameters such as corrosion potential (E_{corr}), corrosion current density (i_{corr}), pitting potential (E_{pit}), protection potential (E_{pp}), and recorrosion current density (i_{rcorr}) are collected

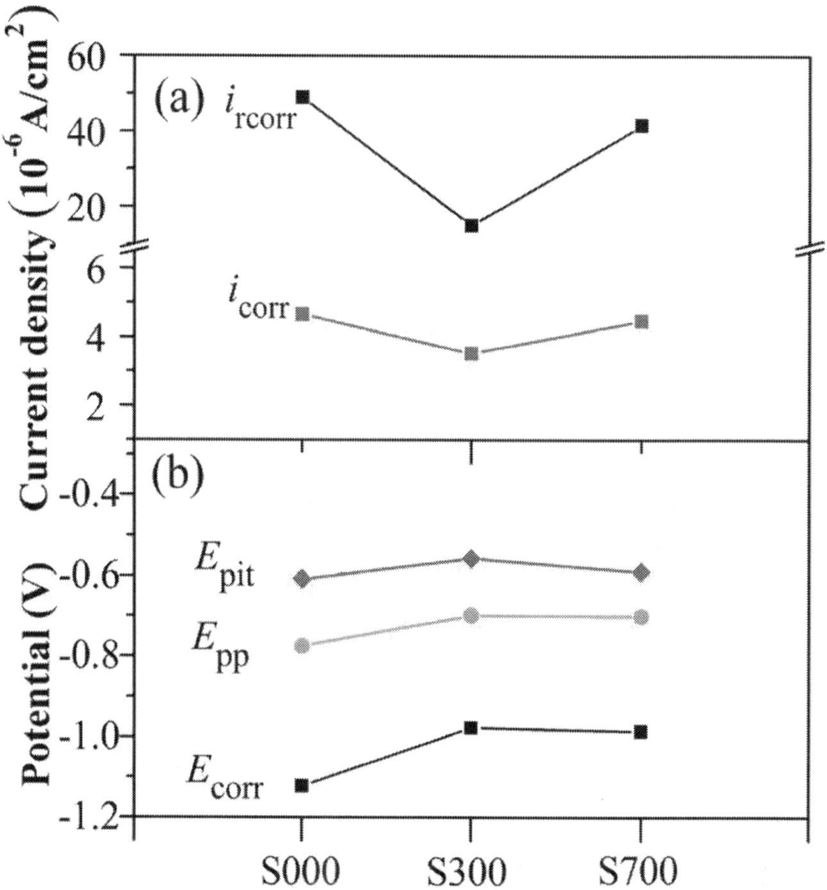

Figure 5. Parameters of the polarization obtained from Figure 4: (**a**) corrosion current density (i_{corr}) and recorrosion current density i_{rcorr}; (**b**) corrosion potential (E_{corr}), pitting potential (E_{pit}) and protection potential (E_{pp}).

in Figure 5. The E_{corr} of the three samples are also in the order S300 > S700 > S000, and i_{corr} is quite the opposite, indicating that the anodic reaction of FSP samples occurs with more difficulty in the electrochemical test. Furthermore, the E_{pit} and E_{pp} of S300 and S700 are more positive than that of S000, and S300 has the highest E_{pit} and E_{pp}, which is consistent with the measured E_{corr}. Here, the concept of protection potential (E_{pp}) represents the potential at which pre-existing pits cease to propagate, that is, all the propagating pits are repassivated [23]. The E_{pit} andE_{pp} are related to pitting corrosion resistance, and the higher E_{pit} and E_{pp} indicate higher resistance to pitting corrosion. The i_{rcorr} represents the current density at which the samples incur corrosion after the repassivation region. The i_{rcorr} of FSP samples is lower than base metal, and the i_{rcorr} of S300 is the lowest, *i.e.*, in the same order of i_{corr}. Combined with the above measured electrochemical parameters, it is found that the corrosion resistance of the three samples is in the order S300 > S700 > S000.

The SEM images of electrochemical corroded surfaces of base metal and FSP samples are shown in Figure 6. The measured samples were polarized until −0.4 V_{SCE} with a scanning rate of 1 mV/s and without backward scanning (the polarization curves are not shown here). Many corrosion pits of 50 μm in diameter appear on the surface of S000; by contrast, there are fewer corrosion pits of smaller size (~20 μm) on the surface of S300 and S700. With further investigation, the pits on the surface of S300 are shown to be much sparser than that of S700. The pits on the base metal are deeper than on the FSP samples, accompanied by obvious cracking.

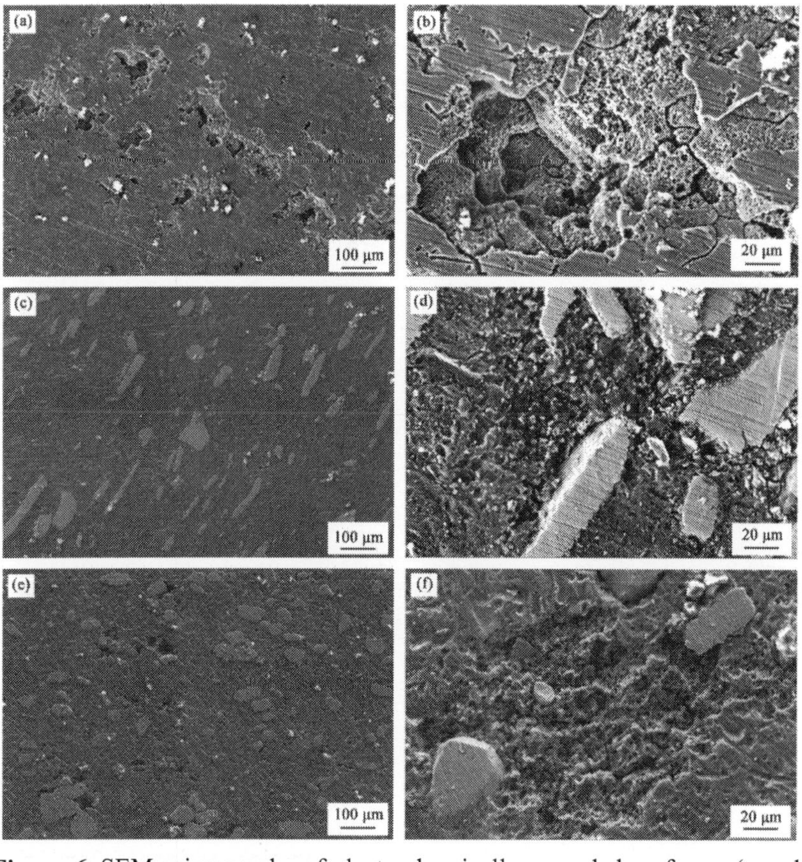

Figure 6. SEM micrographs of electrochemically corroded surfaces: (**a** and **b**) S000; (cand **d**) S300; (**e** and **f**) S700.

Figure 7 shows the weight loss of S000, S300, and S700 samples as a function of sliding time. It is observed that their weight losses increase monotonically with the sliding time, and when the sliding time reaches 7 min, the weight loss of S000 increases more quickly than that of S300 and S700. Overall, the wear resistance of three samples are ranked in the order S300 > S700 > S000, which is consistent with their corrosion resistances (Figure 4 and Figure 6). The SEM micrographs of worn tracks of the three samples are shown in Figure 8. The worn surface of the base metal (S000) shows a large amount of plastic flows. The extent of wear in S300 and S700 is lower than that in S000, *i.e.* their surfaces are smoother and the groove is shallower compared with S000. The higher magnified view of the wear pits shows that the flaking of the base metal is more serious than with the FSP samples, indicating that the worn debris formation is easier at the base metal (Figure 8b,d,f). In addition, there are a lot of fine particles in the size of several microns around the $Fe_{78}Si_9B_{13}$ particles (Figure 8d,f) and there exist white particles that impede the worn pit extend along the sliding direction (Figure 8c), indicating that the added $Fe_{78}Si_9B_{13}$ particles play a role in improving the samples' wear resistance.

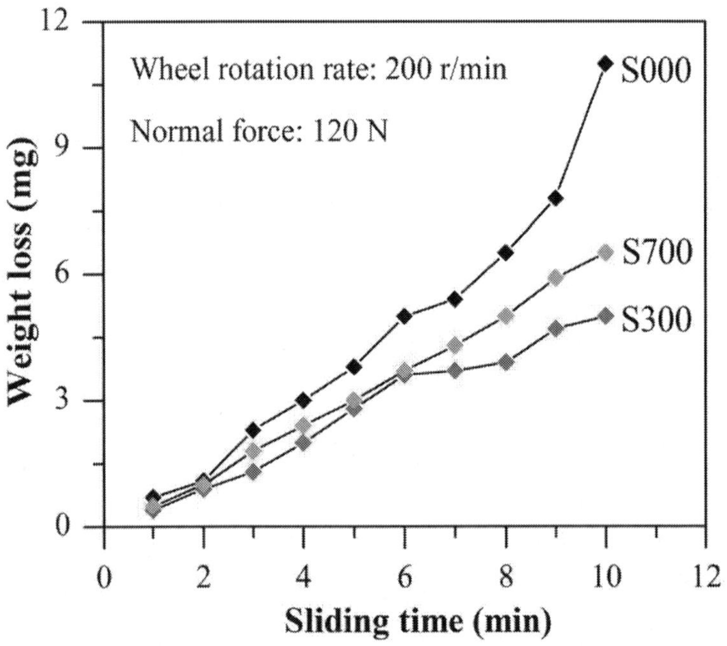

Figure 7. Weight loss of samples as a function of sliding time for base metal (S000) and FSP samples (S300 and S700).

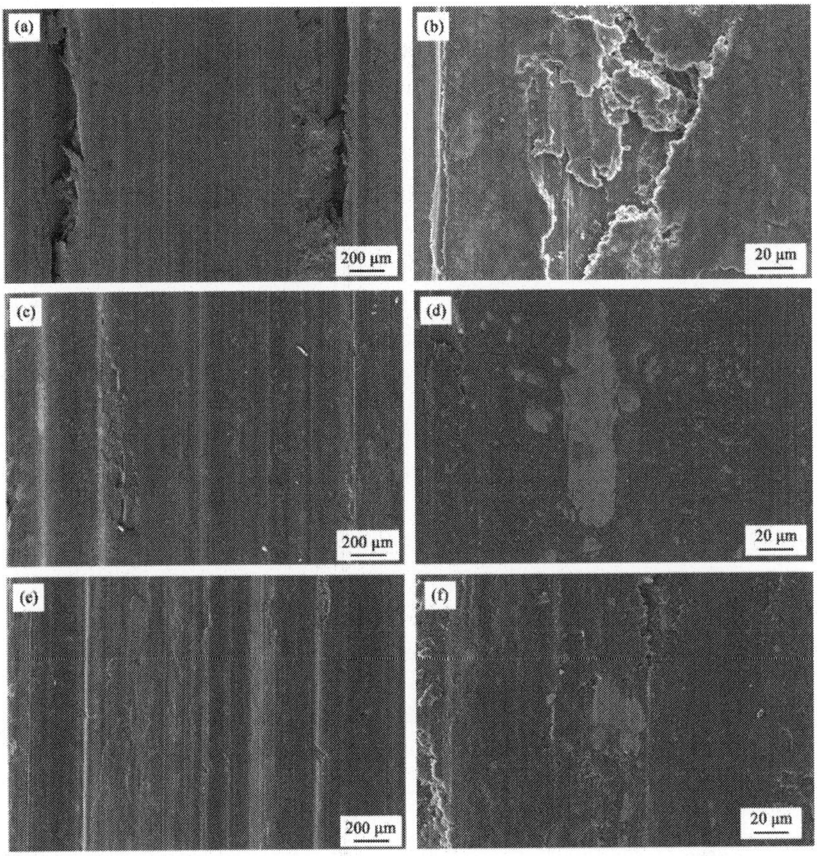

Figure 8. SEM micrographs of worn tracks of samples: (**a** and **b**) S000; (**c** and **d**) S300; (**e**and **f**) S700.

4. DISCUSSION

4.1. Microstructure and Thermal Expansion Behavior of FSP Samples

It is well known that grain refinement in metals and alloys can be realized by friction stir processing (FSP), which was derived based on the principles of friction stir welding (FSW) [9,10] and is similar to the grain refining effect of severe plastic deformation (SPD) [24]. Hence, FSP can be considered one of the variants of the severe plastic deformation (SPD) technologies, which is also discussed in recent reports [25,26]. It has been reported that the SPD treatment accelerates the atomic diffusion inside the processed material, which attributes to the intensive vacancy generation during deformation [27,28,29]. Moreover, the solvus temperature of needle-shape precipitate (β") in 6061 aluminum alloys is about 353 °C, which is lower than the maximal temperature during FSP [30], so the β” can easily dissolve into the matrix during FSP. In addition, the β” mainly contains Mg and Si atoms and the dissolution of Mg and Si into aluminum matrix could increase the lattice constant of α-Al [31], thus it is understood that the lattice parameters of S300 and S700 increase after FSP compared with S000 (Figure 2 inset and Table 1). Furthermore, it is reported that the initial rolling texture is destroyed by the deformation during the FSP [32,33,34]. It is explained that the preferred orientation factor *F* of the (200) plane in α-Al for S300 and S700 is higher than that of S000 (Figure 2 and Table 1).

In the S300 sample, the $Fe_{78}Si_9B_{13}$ particles annealed at 300 °C are in the amorphous state, while the $Fe_{78}Si_9B_{13}$ particles have crystallized completely during the annealing process at 700 °C (Figure 1b). In the amorphous state, the atoms are metastable and in a random order [19], therefore the Si in $Fe_{78}Si_9B_{13}$ is apt to dissolve into the aluminum matrix and form a solid solution during FSP. Furthermore, the dissolution of Si could also induce severe lattice distortion and dislocation, which may lead to the dissolution of more Mg and Si into the aluminum matrix. In the S700 sample, the influence of the $Fe_{78}Si_9B_{13}$ particle annealed at 700 °C on the lattice parameter is weaker due to its crystalline structure. As a consequence, the lattice constant a_0 of α-Al in S300 is larger than that in S700 (Table 1). Meanwhile, compared with S700, the lower *F* of S300 indicates a severer recrystallization and then a severer destruction of rolling texture, which may result in a severer lattice distortion and dislocation in S300 and a higher dissolution degree of Mg and Si. This is also confirmed by the fact that the a_0 of α-Al in S300 is higher than that in S700 (Table 1).

According to the micro-mechanical model [35], the initial compression stress of a sample induced by CTE mismatch of the constituents is relieved with increasing temperature and causes an increment in the coefficient of thermal expansion (CTE). We determined the initial increment of CTE of the three measured samples (Figure 3). Furthermore, the CTEs for the three samples increase at first, which is also related to the dissolution of the β or β” phase, which mainly contains Mg and Si due to the positive contribution of Mg to CTE and a_0 of α-Al solution [31,36,37]. During the cooling process after FSP, the shrinkage degree of base metal was larger than that of the reinforced particles, because the CTE of added particles is less than that of the aluminum matrix alloy, resulting in compressive stress in both the base metal and the added particles [38]. It is understood that the initial CTE of S300 and S700 is lower, and the CTE increment in the beginning stage is higher compared with S000 (Figure 3). Moreover, the maximal CTEs for S300 and S700 are $2.87 \times 10^{-5}\ K^{-1}$ and $3.04 \times 10^{-5}\ K^{-1}$, respectively, and the first maximal of S000 is $2.71 \times 10^{-5}\ K^{-1}$, which is lower than the former values, indicating a higher ability to dissolve Mg for S300 and S700. This is consistent with the a_0 variation of measured samples (Table 1). In addition, the temperature at which the CTE begins to decrease for S300 is higher than that for S700, which is due to increased Mg dissolution into the matrix during FSP for S300, which is also reflected in the data of lattice parameters (Table 1). The decrease of CTE for FSP samples after the maximal CTE is related to the Si dissolution into the matrix, supplied by $Fe_{78}Si_9B_{13}$ particles with a negative contribution to CTE of matrix [39]. Moreover, at 550 °C, the CTE of S300 is lower than that of S000 and S700, which is related to shrinkage of the amorphous $Fe_{78}Si_9B_{13}$ particles in S300 crystallized during heat processing.

4.2. Corrosion and Wear Behavior of FSP Samples

After friction stir processing (FSP), the grain size decreased due to dynamic recrystallization [34], which occurs early on in the alloys after SPD [40,41]. The dynamic recrystallization process has undergone a series of paths [42], which involves the dissolution introduction, dynamic recovery, and continuous dynamic recrystallization. It has been reported that the decrease of grain size can reduce the corrosion rate, that is, it improves the corrosion resistance, which is attributed to an ability of fine-grained materials to passivate more readily and to physical breakdown in the second phase of the intermetallic particles that operated as efficient local cathodes and sites for initiation of localized attack [43,44]. Due to the decrease of grain size by dynamic recrystallization during FSP [33,34], the FSP will improve the corrosion resistance of the aluminum alloy, which is in accordance with Rao's research [45]. On the other hand, in the weld head and its vicinity the residual FSP stresses are either parallel or perpendicular to the weld direction; these are compressive stresses due to the CTE mismatch between the base metal and particles [38,46]. The compressive residual stress can improve the corrosion resistance of the surface by reducing the tendency of cracking [34,47,48]. Hence, S300 and S700 have more positive E_{pit} andE_{pp} than S000 and fewer cracks than S000 (Figure 4, Figure 5 and Figure 6). Because the CTE of $Fe_{78}Si_9B_{13}$ in the amorphous state is less than that in the crystallized state [38], the S300 endured larger compressive stress than S700. Moreover, it has been reported that the amorphous samples have a better corrosion resistance than the crystallized ones [49,50]. Hence, it appears that S300 has more positive E_{corr}, E_{pit}, and E_{pp} than S700 (Figure 4, Figure 5 and Figure 6).

The presence of added particles converting the wear mode of FSP samples from two bodies to three bodies reduces the loss of wear further by acting as a solid lubricant [5,14,51]. Apparently, the fine powders of the size of several microns around the $Fe_{78}Si_9B_{13}$ particles possibly act as lubricants (Figure 8d,f). In addition, the residual stress after FSP is compressive stress [46], and moderate levels of compressive residual stress have beneficial effects on wear resistance by delaying crack initiation and growth [52]. Furthermore, the smaller grain was conducive to formation of a tribological transformed structure (TTS) layer, and produced shorter delamination cracks in the TTS layer than the larger one [53]—that is, the ultrafine-grained microstructures exhibit markedly enhanced wear resistance relative to the coarse-grained counterpart. Consequently, for the particles located at the end of worn pits, the wear loss of S300 and S700 is lower than S000 (Figure 7). It is known that the amorphous $Fe_{78}Si_9B_{13}$ is metastable and has a drastic stress-induced flow at about 400 °C during the heating process [19,54], which possibly accounts for it acting as a better lubricant (Figure 7). As discussed previously, the difference of CTE for added particles leads to a larger compressive stress in S300 than in S700. Hence, it is shown that S300 has better wear resistance than S700 (Figure 7). In other words, amorphous $Fe_{78}Si_9B_{13}$-reinforced particles can improve the wear and corrosion resistance of AA6061 samples.

5. CONCLUSIONS

The microstructure, thermal expansion behavior, corrosion behavior and, wear resistance of AA6061-T6 with $Fe_{78}Si_9B_{13}$glass particles by friction stir processing have been investigated (the base metal is denoted as S000, and the FSP samples with the reinforced particles annealed at 300 °C and 700 °C are denoted as S300 and S700, respectively), and it was found that:

1) By friction stir processing, the $Fe_{78}Si_9B_{13}$ particles are relatively uniform in terms of their dispersion into the Al alloy matrix. The a_0 of α-Al for three samples is in the order S300 > S700 > S000, but the F_{200} for samples is the opposite. During the heating process, the CTEs of the three samples increase firstly and have a maximal value of 2.71×10^{-5} K^{-1}, 2.87×10^{-5} K^{-1}, and 3.04×10^{-5} K^{-1} for S000, S300, and S700, respectively. The CTE variation can be explained by the dissolving of Mg and Si from the β phase and $Fe_{78}Si_9B_{13}$ particles, as well as the CTE mismatch compressive stress.
2) The E_{corr} and E_{pit} of the three samples are also in the order S300 > S700 > S000, while i_{corr} is quite the opposite. The superior corrosion resistance for S300 can be attributed to the reduction of grain size by FSP and the unique metastable structure of amorphous $Fe_{78}Si_9B_{13}$-reinforced particles. Moreover, the corrosion surface of S300 and S700 has little corrosion pits, but no apparent cracking compared to S000.
3) The wear resistance of FSP samples has been improved compared with the base metal; in particular, the weight loss for the samples is in the order S300 < S700 < S000. The flaking of S000 after a wear test is more serious than that in S300 and S700. In addition, the worn surfaces of FSP samples are smoother and the grooves shallower compared with S000. The partial improvement of corrosion and wear resistance is ascribed to the compressive stress in FSP samples due to the CTE mismatch between the

base metal and reinforced particles, and the lubrication effect of $Fe_{78}Si_9B_{13}$ particles also plays a role in improving wear resistance.

ACKNOWLEDGMENTS

This work is supported by the National Natural Science Foundation of China (Grants No. 51171091, 51471099), the Excellent Youth Project of Natural Science Foundation of Shandong (No. JQ201012), and the National Basic Research Program of China (973 Program) (No. 2012CB825702).

AUTHOR CONTRIBUTIONS

Lingyu Guo performed the experiments, analyzed the data, and wrote the initial draft of the manuscript; Weimin Wang designed the project, reviewed and contributed to the final revised manuscript; Yan Liu, Kechang Shen, and Kibuem Kim provided helpful feedback and commented on the manuscript; Min Yang and Chaoqun Song performed the experiments and contributed equipment. All authors read the final paper.

REFERENCES

1. Bakes, H.; Benjamin, D. *Metals Handbook*; ASM international: Metal Park, OH, USA, 1979.
2. Vasudevan, A.K.; Doherty, R.D. *Aluminum Alloys-Contemporary Research and Applications: Contemporary Research and Applications*; Elsevier: Amsterdam, The Netherland, 2012.
3. Ravi, N.; Sastikumar, D.; Subramanian, N.; Nath, A.K.; Masilamani, V. Microhardness and Microstructure Studies on Laser Surface Alloyed Aluminum Alloy with Ni-Cr. *Mater. Manuf. Process.* **2000**, *15*, 395–404.
4. Miracle, D.B. Metal matrix composites–From science to technological significance. *Compos. Sci. Technol.* **2005**, *65*, 2526–2540.
5. Shafiei-Zarghani, A.; Kashani-Bozorg, S.F.; Zarei-Hanzaki, A. Microstructures and mechanical properties of Al/Al_2O_3surface nano-composite layer produced by friction stir processing. *Mater. Sci. Eng. A* **2009**, *500*, 84–91.
6. Budinski, K.G. *Surface Engineering for Wear Resistance*; Prentice Hall: Upper Saddle River, NJ, USA, 1988.
7. Mabhali, L.; Pityana, S.L.; Sacks, N. Laser Surface Alloying of Aluminum (AA1200) with Ni and SiC Powders. *Mater. Manuf. Process.* **2010**, *25*, 1397–1403.

8. Gupta, M.; Mohamed, F.A.; Lavernia, E.J. Solidification behavior of AI-Li-SiCp MMCs processed using variable co-deposition of multi-phase materials. *Mater. Manuf. Process.* **1990**, *5*, 165–196.
9. Mishra, R.S.; Mahoney, M.W.; McFadden, S.X.; Mara, N.A.; Mukherjee, A.K. High strain rate superplasticity in a friction stir processed 7075 Al alloy. *Scripta Mater.* **1999**, *42*, 163–168.
10. Mishra, R.S.; Ma, Z.Y.; Charit, I. Friction stir processing: A novel technique for fabrication of surface composite. *Mater. Sci. Eng. A* **2003**, *341*, 307–310.
11. Ma, Z.Y. Friction stir processing technology: A review. *Metall. Mater. Trans. A* **2008**, *39*, 642–658.
12. Aruri, D.; Adepu, K.; Adepu, K.; Bazavada, K. Wear and mechanical properties of 6061-T6 aluminum alloy surface hybrid composites [(SiC+Gr) and (SiC+Al_2O_3)] fabricated by friction stir processing. *J. Mater. Process. Tech.* **2013**, *2*, 362–369.
13. Clyne, T.W.; Withers, P.J.; Minges, M.L. *An Introduction to Metal Matrix Composites*; Cambridge University Press: Cambridge, UK, 1995.
14. Ahlatci, H.; Ko Er, T.; Candan, E.; Imeno Lu, H. Wear behaviour of Al/(Al_2O_3p+SiCp) hybrid composites. *Tribol. Int.* **2006**, *39*, 213–220.
15. Liu, Z.Y.; Xiao, B.L.; Wang, W.G.; Ma, Z.Y. Tensile strength and electrical conductivity of carbon nanotube reinforced aluminum matrix composites fabricated by powder metallurgy combined with friction stir processing. *J. Mater. Sci. Technol.* **2014**, *30*, 649–655.
16. Makino, A.; Kubota, T.; Chang, C.; Makabe, M.; Inoue, A. FeSiBP bulk metallic glasses with high magnetization and excellent magnetic softness. *J. Magn. Magn. Mater.* **2008**, *320*, 2499–2503.
17. Makino, A.; Kubota, T.; Makabe, M.; Chang, C.T.; Inoue, A. FeSiBP metallic glasses with high glass-forming ability and excellent magnetic properties. *Mater. Sci. Eng. B* **2008**, *148*, 166–170.
18. Inoue, A. Stabilization of metallic supercooled liquid and bulk amorphous alloys. *Acta Mater.* **2000**, *48*, 279–306.
19. Luborsky, F.E. *Amorphous Metallic Alloys*; Butterworth and Co.: London, UK, 1983.
20. Davis, J.R. *Aluminum and Aluminum Alloys*; ASM international: Metal Park, OH, USA, 1993.
21. Zhou, Y. *Materials Analysis and Testing Technology*; Harbin Institute of Technology Press: Harbin, China, 2007.
22. Lotgering, F.K. Topotactical reactions with ferrimagnetic oxides having hexagonal crystal structures—I. *J. Inorg. Nucl. Chem.* **1959**, *9*, 113–123.
23. Wilde, B.E.; Williams, E. The use of current/voltage curves for the study of localized corrosion and passivity breakdown on stainless steels in chloride media. *Electrochim. Acta* **1971**, *16*, 1970–1985.

24. Azushima, A.; Kopp, R.; Korhonen, A.; Yang, D.Y.; Micari, F.; Lahoti, G.D.; Groche, P.; Yanagimoto, J.; Tsuji, N.; Rosochowski, A. Severe plastic deformation (SPD) processes for metals. *CIRP Ann. Manuf. Technol.* **2008**, *57*, 716–735.
25. Nikulin, I.; Malopheyev, S.; Kipelova, A.; Kaibyshev, R. Effect of SPD and friction stir welding on microstructure and mechanical properties of Al-Cu-Mg-Ag sheets. *Mater. Lett.* **2012**, *66*, 311–313.
26. Raj, K.H.; Sharma, R.S.; Singh, P.; Dayal, A. Study of friction stir processing (FSP) and high pressure torsion (HPT) and their effect on mechanical properties. *Proc. Eng.* **2011**, *10*, 2904–2910.
27. Mazilkin, A.A.; Straumal, B.B.; Rabkin, E.; Baretzky, B.; Enders, S.; Protasova, S.G.; Kogtenkova, O.A.; Valiev, R.Z. Softening of nanostructured Al-Zn and Al-Mg alloys after severe plastic deformation. *Acta Mater.* **2006**, *54*, 3933–3939.
28. Straumal, B.B.; Mazilkin, A.A.; Baretzky, B.; Schütz, G.; Rabkin, E.; Valiev, R.Z. Accelerated diffusion and phase transformations in Co-Cu alloys driven by the severe plastic deformation. *Mater. Trans.* **2012**, *53*, 63–71.
29. Straumal, B.B.; Gornakova, A.S.; Mazilkin, A.A.; Fabrichnaya, O.B.; Kriegel, M.J.; Baretzky, B.; Jiang, J.Z.; Dobatkin, S.V. Phase transformations in the severely plastically deformed Zr-Nb alloys. *Mater. Lett.* **2012**, *81*, 225–228.
30. Sato, Y.S.; Kokawa, H.; Enomoto, M.; Jogan, S.; Hashimoto, T. Precipitation sequence in friction stir weld of 6063 aluminum during aging. *Metall. Mater. Trans. A* **1999**, *30*, 3125–3130.
31. Steuwer, A.; Dumont, M.; Peel, M.; Preuss, M.; Withers, P.J. The variation of the unstrained lattice parameter in an AA7010 friction stir weld. *Acta Mater.* **2007**, *55*, 4111–4120.
32. Woo, W.; Choo, H.; Brown, D.W.; Vogel, S.C.; Liaw, P.K.; Feng, Z. Texture analysis of a friction stir processed 6061-T6 aluminum alloy using neutron diffraction. *Acta Mater.* **2006**, *54*, 3871–3882.
33. El-Danaf, E.A.; El-Rayes, M.M.; Soliman, M.S. Friction stir processing: An effective technique to refine grain structure and enhance ductility. *Mater. Design* **2010**, *31*, 1231–1236.
34. Woo, W.; Balogh, L.; Ungár, T.; Choo, H.; Feng, Z. Grain structure and dislocation density measurements in a friction-stir welded aluminum alloy using X-ray peak profile analysis. *Mater. Sci. Eng. A* **2008**, *498*, 308–313.
35. Huber, T.; Degischer, H.P.; Lefranc, G.; Schmitt, T. Thermal expansion studies on aluminium-matrix composites with different reinforcement architecture of SiC particles. *Compos. Sci. Technol.* **2006**, *66*, 2206–2217.
36. Uju, W.A.; Oguocha, I.N.A. A study of thermal expansion of Al-Mg alloy composites containing fly ash. *Mater. Design***2012**, *33*, 503–509.

37. Berry, C.R. Effect of internal strains on linear expansion, X-Ray lattice constant, and density of crystals. *J. Appl. Phys.* **1953**, *24*, 658–659.
38. Ma, H.J.; Zhang, J.T.; Li, G.H.; Zhang, W.X.; Wang, W.M. Effect of Zr on the thermal stability and magnetic properties of $Fe_{78}Si_9B_{13}$ glassy alloy. *J. Alloy. Compd.* **2010**, *501*, 227–232.
39. Zhang, Q.; Wu, G.; Jiang, L.; Chen, G. Thermal expansion and dimensional stability of Al-Si matrix composite reinforced with high content SiC. *Mater. Chem. Phys.* **2003**, *82*, 780–785.
40. Straumal, B.B.; Protasova, S.G.; Mazilkin, A.A.; Rabkin, E.; Goll, D.; Schütz, G.; Baretzky, B.; Valiev, R.Z. Deformation-driven formation of equilibrium phases in the Cu–Ni alloys. *J. Mater. Sci.* **2012**, *47*, 360–367.
41. Straumal, B.B.; Dobatkin, S.V.; Rodin, A.O.; Protasova, S.G.; Mazilkin, A.A.; Goll, D.; Baretzky, B. Structure and properties of nanograined Fe-C alloys after severe plastic deformation. *Adv. Eng. Mater.* **2011**, *13*, 463–469.
42. Su, J.Q.; Nelson, T.W.; Mishra, R.; Mahoney, M. Microstructural investigation of friction stir welded 7050-T651 aluminium. *Acta Mater.* **2003**, *51*, 713–729.
43. Ralston, K.D.; Birbilis, N.; Davies, C.H.J. Revealing the relationship between grain size and corrosion rate of metals.*Scripta Mater.* **2010**, *63*, 1201–1204.
44. Ralston, K.D.; Birbilis, N. Effect of grain size on corrosion: A review. *Corrosion* **2010**, *66*, 1201–1204.
45. Rao, A.G.; Katkar, V.A.; Gunasekaran, G.; Deshmukh, V.P.; Prabhu, N.; Kashyap, B.P. Effect of multipass friction stir processing on corrosion resistance of hypereutectic Al-30Si alloy. *Corros. Sci.* **2014**, *83*, 198–208.
46. Staron, P.; KoçAk, M.; Williams, S.; Wescott, A. Residual stress in friction stir-welded Al sheets. *Physica B* **2004**, *350*, E491–E493.
47. Van Boven, G.; Chen, W.; Rogge, R. The role of residual stress in neutral pH stress corrosion cracking of pipeline steels. Part I: Pitting and cracking occurrence. *Acta Mater.* **2007**, *55*, 29–42.
48. Prevey, P.; Cammett, J. The influence of surface enhancement by low plasticity burnishing on the corrosion fatigue performance of AA7075-T6. *Int. J. Fatigue* **2004**, *26*, 975–982.
49. Tam, M.K.; Shek, C.H. Crystallization and corrosion resistance of $Cu_{50}Zr_{45}Al_5$ bulk amorphous alloy. *Mater. Chem. Phys.* **2006**, *100*, 34–37.
50. Naka, M.; Hashimoto, K.; Masumoto, T. Corrosion behavior of amorphous and crystalline $Cu_{50}Ti_{50}$ and Cu_{50} Zr_{50}alloys. *J. Non-Cryst. Solids* **1978**, *30*, 29–36.
51. Gurcan, A.B.; Baker, T.N. Wear behaviour of AA6061 aluminium alloy and its composites. *Wear* **1995**, *188*, 185–191.

52. Mohammadzadeh, R.; Akbari, A.; Drouet, M. Microstructure and wear properties of AISI M2 tool steel on RF plasma nitriding at different N_2–H_2 gas compositions. *Surf. Coat. Technol.* **2014**, *258*, 566–573.
53. Li, J.; Lu, Y.; Zhang, H.; Xin, L. Effect of grain size and hardness on fretting wear behavior of Inconel 600 alloys. *Tribol. Int.* **2015**, *81*, 215–222.
54. Niu, Y.C.; Bian, X.F.; Wang, W.M.; Jin, S.F.; Liu, X.J.; Zhang, J.Y.; Qin, G.L. The use of flow behavior and thermal expansion to monitor structural change of amorphous $Fe_{78}Si_9B_{13}$ ribbon. *J. Non-Cryst. Solids* **2005**, *351*, 3854–3860.

Chapter 4

Wear Resistance and Indentation Behavior of Equiatomic Superelastic TiNi and 60NiTi

Rabin Neupane, Zoheir Farhat

Department of Process Engineering and Applied Science, Materials Engineering Program, Dalhousie University, Halifax, Canada

ABSTRACT

Indentation and reciprocating wear tests are carried out to study dent and wear resistance of superelastic Ti-Ni alloys. The effect of loading rate on the superelastic behavior of TiNi under indentation loading is investigated and compared to a new generation of shape memory alloys, i.e., 60NiTi. Only limited amount of work has been done to investigate the dependency of superelasticity on loading rate of TiNi under localized compressive loads, but much work is directed towards understanding the effect of strain rate on tensile properties. Understanding the superelastic behavior helps to employ superelastic alloys in applications where high impact loading is expected as in bearings and gears. In the present study, it is found that dent resistance of Ti-Ni alloy is not significantly affected by loading rate (within the employed loading conditions). It has also been found that new-generation 60NiTi alloy exhibits superior wear and dent resistance, as well as higher hardness compared to equiatomic TiNi.

Keywords: TiNi, 60NiTi, Superelasticity, Dent Resistance, Wear Resistance

1. INTRODUCTION

TiNi shape memory alloys are widely known for shape memory and superelastic effects, which occur due to reversible martensitic transformation that can be induced either thermally or mechanically [1] [2] . TiNi exhibits superelastic effect through stress-induced martensitic transformation (i.e., phase transformation) between parent austenite phase and martensite phase, leading to high recoverable strain during deformation. During loading, the parent austenite phase transforms to twinned martensite followed by detwinning of martensite, i.e., reorientation of martensitic variants, accompanied by large strain. The reverse transformation from martensite to austenite occurs upon unloading, accompanied by large recoverable strain [2] .

TiNi is widely used in a variety of applications in automotive, aerospace, and biomedical industries due to their high corrosion resistance, good fatigue properties, high mechanical damping and good ductility [2] -[5] . Recent studies have shown that superelastic TiNi has superior wear and dent resistance compared to other conventional materials [3] [6] [7] . There are currently extensive efforts to develop bearings and gears using superelastic TiNi alloys [8] [9] . Conventional bearing materials, such as AISI 52100, AISI M-50, AISI 440C steels and carbon steels, are susceptible to denting and wear during high loading [10] . It is expected that new high precision bearings made from superelastic TiNi alloy may provide superior wear and dent resistance during high impact loading conditions. In the case of conventional materials, wear and dent resistance depend on their mechanical properties such as strength, hardness, toughness, ductility, work hardening and crystallography texture [3] -[5] . However, for superelastic TiNi, in addition to mechanical properties, stress-induced martensitic transformation plays a major role in the observed wear and dent behavior [3] [10] [11] .

The micromechanical and tribological characteristics of superelastic TiNi are studied by Farhat et al. [11] . They performed wear tests on superelastic TiNi and compared to Ti, Ni and AISI 304 steel using pin-on-disc configuration. Amongst all the materials tested, TiNi alloy showed about 30 and 10 times higher wear resistance than pure Ti and Ni, respectively. Li et al. [12] studied the sliding wear behavior of Ti-50.3 at %Ni alloy and 2Cr13 steel. They found that TiNi alloy had much higher wear resistance than steel, although it had lower hardness. The superior wear resistance of superelastic TiNi alloy was attributed to low E/H and high elastic recovery ratio [3] [11] . Lin et al. [13] studied the effect of various applied loads on wear characteristics of TiNi shape memory alloy. It was found that normal load, sliding distance, and sliding speed had considerable influences on superelasticity of TiNi alloy. Their experimental results showed that during sliding wear tests, weight loss increased with increasing load according to Archard's law [14] . They suggested that under low loads, the material was superelastic while at high loads superelasticity diminished. Recent studies on 60NiTi have shown a unique combination of high hardness, low elastic modulus and superior elasticity making it a potential candidate for bearing and gear applications [7] [15] [16] . Static indentation load capacity tests on 60NiTi were conducted by DellaCorte et al. [15] . They found that 60NiTi surfaces could resist higher load levels before permanent deformation than conventional bearing steels, such as AISI 440C, AISI M-50, AISI 52100, and REX 20. They attributed this behavior to increase in contact area and reductions in peak and mean contact stresses.

Furthermore, in addition to dent resistance, loading rate is also of significant importance to applications where denting is expected. Substantial amount of work has been done towards understanding the effect of strain rate under tensile properties, i.e., uniaxial loading. Recent studies on strain rate dependency on superelasticity of TiNi under tensile loading conditions have shown that increasing strain rate leads to shortening of the stress plateau and increasing the austenite to martensite transition stress during the deformation process [17] -[21] . The rise in transition stress was related to heat accumulation in the deformation zone during uniaxial loading [17] [18] [21] -[23] . It has also been found that the transition stress increases with increasing strain rate whereas recoverable strain decreases with increasing strain rate [24] . On the other hand, only limited work has been done to understand the effect of loading rate on superelastic TiNi under indentation loading, i.e., under localized compressive loads. Several recent studies have shown a relationship between loading rate and superelasticity under indentation conditions [10] [25] [26] . It was found that increasing loading rate partially diminished the martensite transformation in superelastic TiNi alloy. In other words, under high loading rates, superelastic TiNi alloy behaves similar to conventional materials. It has been proposed that the drop in superelasticity under high loading rates may be due to retardation of stress-induced martensite transformation [10] . Much work is needed to fully understand the effect of loading rate on superelastic behavior of TiNi under indentation loading conditions. Hence, the objective of the present work is to develop an understanding of the dent and wear behavior of equiatomic TiNi and compare to a new generation of shape memory alloys, i.e., 60NiTi.

2. MATERIALS AND METHODS

In order to investigate the superelastic effect of TiNi and 60NiTi under indentation and wear conditions, indentation and reciprocating wear tests were performed using a multi-function nano-mechanical/tribometer system using Berkovich indenter and tungsten carbide (WC) ball, respectively. A flat plate, 458 × 97 × 1 mm of superelastic equiatomic TiNi alloy was obtained from Johnson Matthey Inc. and cut into squares of 15 × 15 × 1 mm and mounted in

Bakelite. The TiNi sheets were oxide free and in a semi polished conditions. The chemical composition (weight %) of the as-received superelastic TiNi alloy is given in Table 1. 60NiTi aged, annealed, and solution treated blocks having dimensions of 45 × 45 × 5 mm were prepared at National Aeronautics and Space Administration (NASA), Glenn Research Center, Cleveland, Ohio. The blocks of 60NiTi were cut into squares of 14 × 14 × 5 mm and mounted in Bakelite. The chemical compositions (weight %) of the as-received 60NiTi under aged, annealed, and solution treated conditions are given inTable 2. Specimens were ground using 240, 320, 400 and 600 grit SiC abrasive paper followed by polishing using 1, 0.3 and 0.05 μm aluminum oxide abrasive. Crystal structures and phases of equiatomic TiNi and aged, annealed, and solution treated 60NiTi were determined from x-ray diffraction patterns. X-ray diffraction (XRD) tests were carried out using a high speed Bruker D8 Advance XRD system with CuKα radiation, tube current of 40 mA, tube voltage of 40 kV and radiation wavelength of 1.54 Å. The diffraction patterns were analyzed and compared to Powder Diffraction Files (PDF) database. The strongest peaks in all alloys are austenite peaks, which indicate that all alloys are austenitic at room temperature, hence, exhibiting superelastic behavior.

Indentation tests under varying loading rates and normal loads were conducted on TiNi and 60NiTi alloys using a nanoindentation system. The instrument uses a Berkovich diamond pyramid indenter with an angle of 65.3° between the tip axis and the faces of the triangular pyramid. The displacement of indentation and the load are measured independently with resolutions of 0.03 nm and 0.1 μN. Hardness and elastic modulus of TiNi and 60NiTi were calculated from depth versus load curves using Oliver and Pharr's method [27] . The hardness is calculated according to,

$$H = \frac{P}{24.5h_c^2} \tag{1}$$

where, h_c is the contact depth of penetration and can be calculated from,

$$h_t = h_c + \left[\frac{2(\pi - 2)}{\pi}\right]\frac{P_{max}}{dP/dh} \tag{2}$$

where, h_t is the total penetration depth, P_{max} is the maximum load and dP/dh is the slope of the unloading curve. The elastic modulus of the system is given by,

$$E^* = \frac{dP}{dh}\frac{1}{2}\frac{\sqrt{\pi}}{\sqrt{A}} \quad (3)$$

where,

$$\frac{1}{E^*} = \frac{(1-\nu^2)}{E} + \frac{(1-\nu_i^2)}{E_i} \quad (4)$$

where, E and ν are the elastic modulus and Poisson's ratio for the specimen and E_i and ν_i are the same parameters for the Berkovich indenter. In this study E_i and ν_i were taken as 1141 GPa and 0.07, respectively. The loading rate and maximum load combinations were varied according to Table 3, i.e., a total of 63 tests for equiatomic TiNi, and 189 tests for 60NiTi (aged, annealed and solution treated) were performed and each test was repeated twice. Indentation tests showed excellent repeatability of not more than 5% variation. Loading and unloading (load versus depth) curves for different maximum loads were generated as a function of loading rate.

Table 1. Chemical composition of equiatomic TiNi (wt%).

Elements	Ni	Ti	Fe	C	O	H	Others
Wt%	55.99	43.68	0.05	<0.05	0.0216	<0.005	<0.20

Table 2. Chemical composition of aged, annealed, and solution treated 60NiTi (wt%).

Elements	Ni	Ti	Fe	Al	Cu	Ca	Others
Aged 60NiTi wt%	57.22	40.07	0.009	0.017	0.017	0.012	<0.10
Annealed 60NiTi wt%	60.23	41.79	0.007	0.019	0.014	0.009	<0.10
Solution treated 60NiTi wt%	61.52	43.16	0.009	0.021	0.011	0.015	<0.10

Table 3. Indentation test conditions.

Loading rate (mN/min)	300, 600, 1800, 3000, 4200, 6000, 18000
Max load (mN)	40, 70, 100, 200, 250, 300, 350, 400, 450

To investigate the tribological properties of equiatomic TiNi and 60NiTi, reciprocating wear tests were conducted. Dry reciprocating wear tests were performed using a Universal Mirco-Tribometer (UMT) system. This test method involves a ball upper counter-face that slides against a flat lower specimen of polished TiNi and 60NiTi in a linear back and forth sliding motion, having a stroke length of 5.03 mm. The load on the flat specimen is applied vertically downward with a motor driven carriage that uses a load sensor for feedback to maintain a constant load. The specimens were securely fastened inside the wear chamber. All the tests were conducted at room temperature with relative humidity of 40% - 55%. A 6.3 mm diameter tungsten carbide (WC) ball having a hardness of HRA 92 was used as a counter-face material. A WC ball was selected because it has much higher hardness than superelastic TiNi and 60NiTi. This minimizes wear of the counter-face and prevents shape change, which can lead to change in the contact geometry of the tribo-system, hence, change in mean pressure during testing. The tungsten carbide ball was mounted inside a ball holder which is attached directly to a suspension system and a load sensor that controls and records forces during the test. The instantaneous values of calibrated normal load (Fz) and tangential load (Fx) were measured and continuously recorded using a data acquisition system. The system automatically computes the variation of the coefficient of friction (COF = Fx/Fz) with time. Wear tests for equiatomic superleastic TiNi and 60NiTi (aged, annealed and solution treated) were performed under normal loads of 20, 40, and 60 N with varying reciprocating frequencies from 5 to 20 Hz and time intervals between 30 to 180 minutes. The weight of specimens was measured before and after each test to determine individual weight-loss at 30 minutes intervals. The test method is based on ASTM standard G133- 05(2010) [28] , i.e. standard test method for reciprocating sliding wear. After reciprocating wear tests, worn surfaces, tungsten carbide ball surface, and wear debris were examined using Hitachi S-4700 cold field Scanning electron microscopy. An Oxford® X-Sight 7200 Energy Dispersive X-ray Spectroscopy (EDS) was used for chemical analysis of wear tracks.

3. RESULTS AND DISCUSSION

Optical micrographs of equiatomic TiNi and aged, annealed, and solution treated 60NiTi are shown in Figures 1(a)-(d). Superelastic TiNi alloy (Figure 1(a)) shows regular equiaxed grains. The grain size appears to be uniform with an average size of 20 μm. On the other hand, aged, annealed, and solution treated 60NiTi show somewhat irregular and less uniform grain structure (Figures 1(b)-(d)). The average grain size appears to be 100 μm.

60NiTi microstructure can be very complex depending on thermal treatment [1] . Depending upon aging temperature and time, metastable intermetallic phases (precipitates) such as Ti_2Ni, $TiNi_3$, Ti_2Ni_3 and Ti_3Ni_4 may form. Ti_3Ni_4 is the stable phase at lower aging temperature (680°C) and shorter time which changes to Ti_2Ni_3 upon increasing aging temperature (750°C) and extended time. $TiNi_3$, on the other hand, appears at higher aging temperature (800°C) and longer aging time. Whether the precipitates are observed or not by optical metallography would depend on their size and coherency.

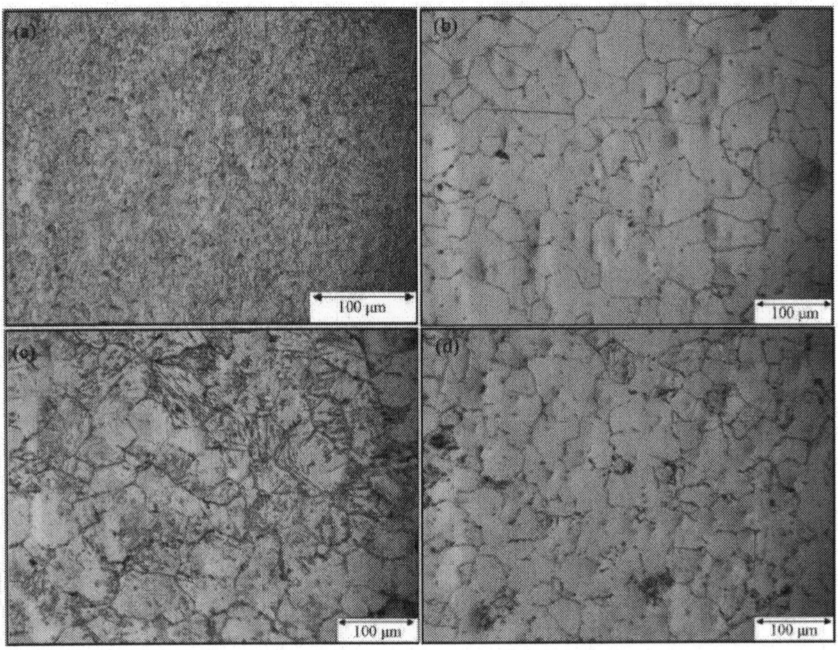

Figure 1. Optical micrographs: (a) equiatomic TiNi; (b) 60NiTi (aged); (c) 60NiTi (annealed); (d) 60NiTi (solution treated).

To evaluate the effect of loading rate on the deformation characteristics of superelastic TiNi and 60NiTi, loading-unloading (load versus depth) curves were generated as a function of maximum load and loading rate. Representative force-displacement (penetration depth of indenter) curves showing loading-unloading generated under 3000 mN/min loading rate and different maximum loads (for equiatomic TiNi as representative curves) using Berkovich indenter are given in Figure 2. The curves in the figure reveal typical load versus displacement patterns, i.e., a rise in indentation depth with increasing normal load. The flat part of the curves at the max load and at the bottom part of the unloading curves is allowed to compensate for creep and thermal drift effects. All the curves generated using different maximum loads overlap, as expected. The curves show high recoverable deformation and relatively small permanent deformation upon unloading. An average of over 60% recovery is observed as a result of the elastic and superelastic behavior of the alloys. It is also evident that even the indentation tests performed under the lowest load of 40mN have shown residual plastic deformation, which is expected for sharp indenter. The curves show the variation in hysteresis size as a function of maximum load, i.e., increases with maximum load.

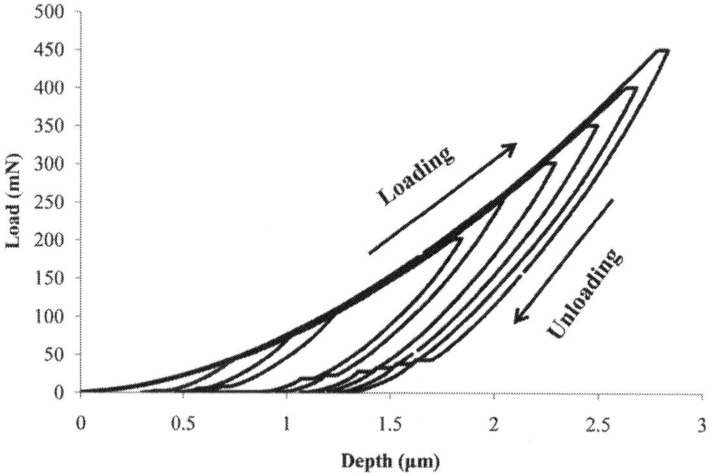

Figure 2. Indentation curves performed using 3000 mN/min loading rate with varying maximum loads for equiatomic TiNi.

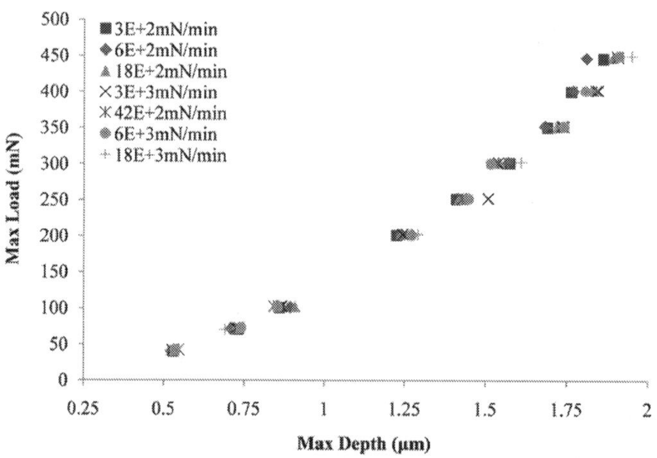

Figure 3. Max depth versus max load as a function of various loading rates for 60NiTi (aged).

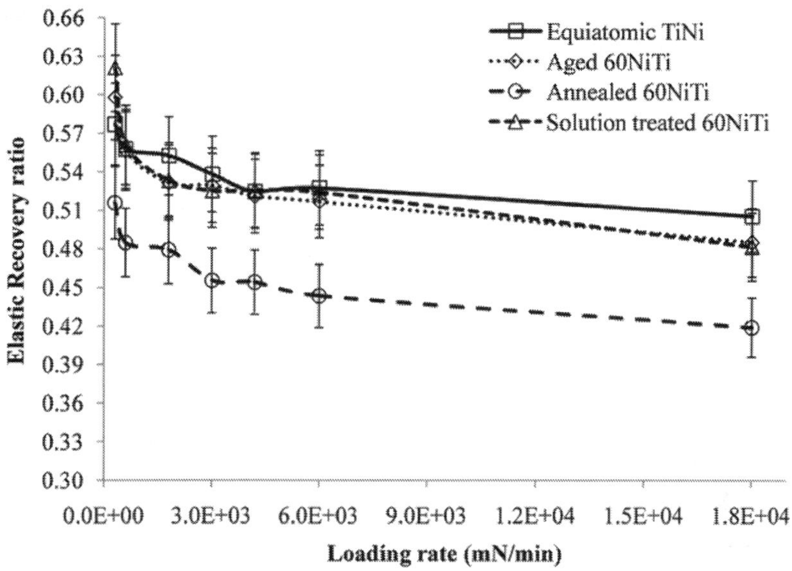

Figure 4. Elastic recovery ratio as a function of loading rate.

Table 4. Mechanical properties of equiatomic TiNi, 60NiTi (aged, annealed and solution treated) and AISI 304 stainless steel.

Properties	Hardness (H) (GPa) (±0.2)	Young's modulus (E) (GPa) (±4)	Elastic recovery ratio	E/H
Equiatomic TiNi	3.2	50	0.53 ± 0.02	15
60NiTi (aged)	9.7	122	0.53 ± 0.02	12
60NiTi (annealed)	5.3	93	0.46 ± 0.02	17
60NiTi (solution treated)	9.3	113	0.53 ± 0.02	12
AISI 304 stainless steel [11]	2.7	229	0.070 ± 0.003	85

Using data collected from indentation tests, maximum load versus maximum depth charts can be generated. Maximum load versus maximum depth values were plotted for aged 60NiTi as representative curve (Figure 3) to assess the effect of loading rate on the deformation characteristics of superelastic alloys. According to Figure 3, the maximum depth of penetration seems to increase linearly with load for all loading rates. Data generated using different loading rates overlap with some scatter at high maximum loads. Similar trend was observed for all materials tested. This behavior indicates that the superelastic behavior of TiNi and 60NiTi, under the indentation conditions used in the present work is independent of loading rate. To further confirm this behavior, elastic recovery ratios were calculated for all loading-unloading curves. The elastic recovery ratio indicates the extent of elastic recovery during deformation and is calculated as the total area under the unloading curve over the total area under the loading curve. The elastic recovery ratio represents a better technique in assessing the effect of loading rate on the superelastic behavior, as this method takes the whole loading-unloading curve into consideration, as opposed to Figure 3 where only the maximum load and maximum depth were considered. The average elastic recovery ratio is plotted as a function of loading rate, as shown in Figure 4. The error bars indicate the scatter of the elastic recovery ratio calculated for various maximum loads at different loading rates. The average elastic recovery ratio for the equiatomic superelastic TiNi is 0.53, which is over 7 times that of AISI 304 stainless steel (0.07) [11] although the hardness values of both materials are similar. The average elastic recovery ratios for 60NiTi (aged, annealed, and solution treated) are 0.53, 0.46 and 0.53, respectively. It is believed that high elastic recovery ratio contributes to improved dent and wear resistance. That is, superelastic TiNi and 60NiTi have the capacity to accommodate high denting loads without permanently deforming. A closer examination of Figure 4 revels that the elastic recovery ratio is high at low loading rates and drops to steady-state, as loading rate increases. An approximate 5% - 10% drop is observed in the elastic recovery ratio, which represents a mild drop in superelasticity as the loading rate is increased. This behavior might be attributed to retardation of the stress-induced martensitic transformation, i.e., the retardation of twin boundary mobility upon loading at high loading rates. Therefore, more plastic deformation takes place under the indenter tip, which leads to less recovery upon unloading. The mild dependency of superelasticity on loading rate (Figure 4) is not evident from Figure 2 andFigure 3, as the elastic recovery ratio represents a better measure of the amount of recoverable deformation.

Table 4 summarizes the mechanical properties of equiatomic TiNi, 60NiTi (aged, annealed and solution treated) and steel. The elastic recovery ratio given in Table 4 is the average value calculated for all the loading rates. It is interesting to note that the average elastic recovery for equiatomic TiNi and 60NiTi under different treatments are similar (except for annealed 60NiTi). The lower elastic recovery of the annealed 60NiTi is attributed to its higher E/H ratio. Furthermore, it is evident from Table 4 that aged and solution treated 60NiTi exhibit hardness values of about three times that of equiatomic TiNi, while their E/H values are lower than that of equiatomic TiNi. It is established in the open literature [3] [11] that superelastic materials exhibit low E/H ratio. Therefore, it can be concluded that aged and solution treated 60NiTi have enhanced superelastic behavior and superior hardness over equiatomic TiNi.

Reciprocating wear tests were performed under varying normal loads, frequencies and time intervals to study the wear behavior of equiatomic TiNi and aged, annealed, and solution treated 60NiTi. The cumulative weight loss versus sliding distance data was plotted and representative curves of equiatomic TiNi are shown in Figure 5. The typical variation in wear rates associated with these kinds of tests is less than 10%. All the curves are characterized by two wear regimes. Initially the slope of the curve is high, i.e. high wear rate which is believed to be due to a "break-in" period as seen in Figure 5. The slope of the curve decreases to a lower steady state value, i.e. lower wear rate after a short sliding distance. In this break-in period, ploughing and polishing action occurs. Here, hard asperities of the WC ball plough through the softer TiNi surface and gradually remove TiNi asperities by polishing action accompanied by a drop in weight loss. Under the applied normal loads, as soon as the WC counterface ball comes in contact with the TiNi surface and sliding begins, it deforms the specimen surface and alters its roughness. The figure also shows that the weight loss increases with the increase in applied normal load for a constant reciprocating frequency of 10 Hz (5, 15 and 20 Hz also show the same trend). This behavior is attributed to the fact that higher applied normal loads promote higher levels of deformation leading to higher material removal rate. Similar behavior was observed for 60NiTi (aged, annealed and solution treated).

Typical variations of steady state wear rate as a function of frequency and normal load are given inFigure 6. Each data point in the diagram represents the steady-state wear rate for a given frequency and normal load.

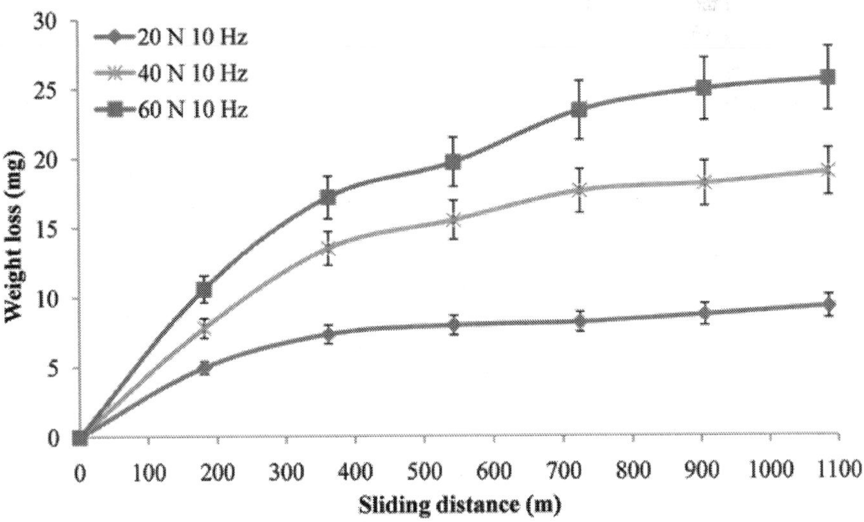

Figure 5. Cumulative weight loss versus sliding distance for different loads for equiatomic TiNi.

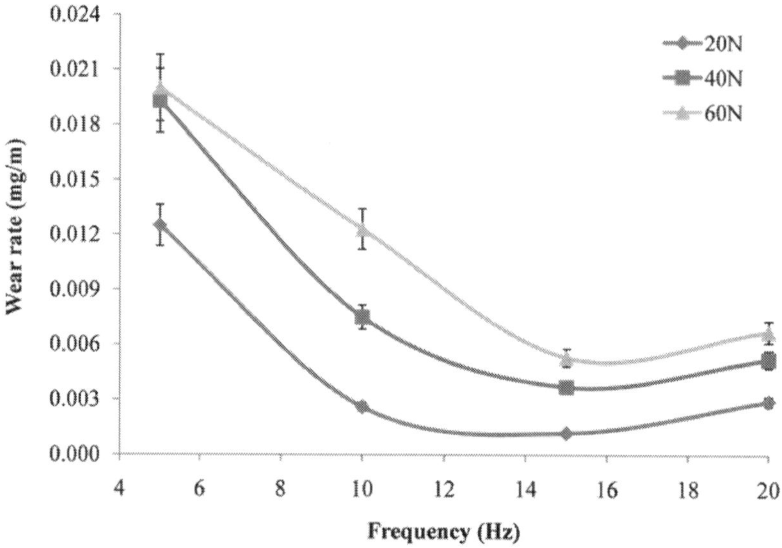

Figure 6. Steady state wear rate as a function of frequency for equiatomic TiNi.

These data points are calculated from the corresponding slope of the steady state region (see Figure 5). Normal load and frequency show significant influence on wear resistance as seen in Figure 5 andFigure 6. The wear rate in Figure 6 is controlled by two effects occurring simultaneously. The first effect is due to softening of the superelastic alloys at high frequency as a result of heating and ultimately leading to higher wear. The second effect is due to rise in the austenite-to-martensite transition stress with temperature [29] , which leads to increased wear resistance. The above discussed effects determine the observed wear rate. A drop in wear rate is clearly seen with increasing frequency up to 15 Hz followed by a mild rise to 20 Hz. In the frequency range 5 to 15 Hz, hardening effect due to rise in austenite-to-martensite transition stress as a result of increased temperature is dominant. However, when the frequency is raised over 15 Hz, the softening effect due to heating seems to be more dominant. Furthermore, the steady-state wear rate increases with increasing normal load.

During the reciprocating wear tests, the instantaneous values of calibrated normal force (Fz) and tangential force (Fx) were measured and the coefficient of friction (COF) was calculated. The COF is plotted as a function of sliding distance for different loads. Representative curves of equiatomic TiNi are shown in Figure 7. The basic shape of the friction curve remains similar under all testing conditions and materials. COF versus sliding distance curves is characterized by two friction regimes. Initially, the coefficient of friction increases until it reaches a maximum steady-state value. The COF becomes independent of sliding distance at steady-state region.

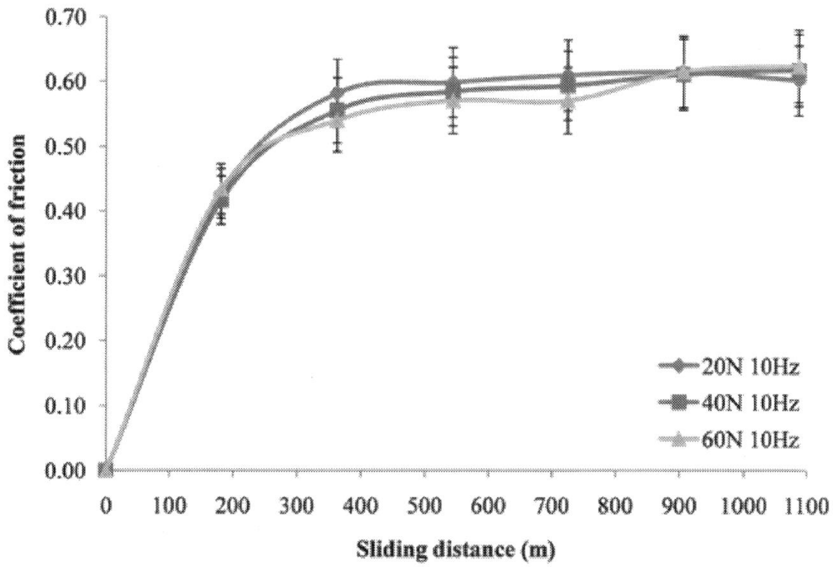

Figure 7. Coefficient of friction versus sliding distance for different loads for equiatomic TiNi.

Table 5. Steady-state COF of equiatomic TiNi and 60NiTi (aged, annealed and solution treated).

Frequency (Hz)	Equiatomic TiNi	60NiTi (aged)	60NiTi (annealed)	60NiTi (solution treated)
5	0.60	0.59	0.58	0.54
10	0.58	0.56	0.54	0.53
15	0.51	0.52	0.52	0.49
20	0.45	0.46	0.46	0.43

Table 5 gives the steady-state coefficient of friction as a function of frequency for equiatomic TiNi and60NiTi (aged, annealed and solution treated). According to Table 5, the COF drops as the reciprocating frequency rises for all the employed normal loads. It is well established that the COF is controlled by both surface and bulk properties in accordance with the general rule [30] :

$$\mathrm{COF} = \tau_s / H_b$$

where, τ_s is surface shear strength and H_b is the bulk hardness. It can be argued that the drop in coefficient of friction with frequency is dominated by surface softening due to heating, which gives rise to a corresponding drop in surface shear strength. Hence, the COF seems to vary only with heating effects. On the other hand, it is found that there is no significant change in the COF, with normal load and each COF value given in Table 5 is an average of the COF values measured using 20, 40 and 60 N loads. As seen in Table 5, the values of the steady-state COF for equiatomic TiNi and 60NiTi (aged, annealed and solution treated) range between 0.43 and 0.60. Furthermore, the average coefficient of friction over the frequency range (5 - 20 Hz) are 0.54, 0.53, 0.53 and 0.50, respectively for equiatomic TiNi, 60NiTi (aged), 60NiTi (annealed), and 60NiTi (solution treated).

Wear data for equiatomic TiNi and 60NiTi (aged, annealed and solution treated) can be described by Archards-type law for sliding wear and the specific wear rate can be determined using Archard's equation, $V = kFs$, where, V is wear volume, k is specific wear rate, F is normal load, and s is sliding distance. Wear volume is calculated by dividing the measured weight loss (for 10 Hz frequency) by the density of TiNi (6.45 g/cc) and 60NiTi (6.7 g/cc). Table 6 summarizes the specific wear rates of equiatomic TiNi and 60NiTi (aged, annealed and solution treated). Examination of Table 6 shows that aged and solution treated 60NiTi exhibit about 40% improvement in wear resistance over equiatomic TiNi. This enhancement in wear resistance is due to the higher hardness of aged and solution treated 60NiTi compared to equiatomic TiNi (see Table 4). The specific wear rates for TiNi and 60NiTi reported in Table 6 are lower than those reported for many wear resistant materials [31] [32] . Equiatomic superelastic TiNi and 60NiTi show lower specific wear rate compared to medium carbon steel (10^{-4} mm^3/Nm), stainless steel (10^{-4} mm^3/Nm), high carbon and tool steels (range 10^{-4} mm^3/Nm to 10^{-5} mm^3/Nm) [31] and, hardened AISI 52100 bearing steel (range $9.6 \times 10^{-5} mm^3/Nm$ to 1.6×10^{-4} mm^3/Nm) [32] . Here, hardened AISI 52100 bearing steel data was generated using ball-on-disc test. However, in the absence of data generated using reciprocating sliding wear, it is reasonable to state, based on the available data, that equiatomic TiNi and 60NiTi exhibit superior wear resistance.

Table 6. Specific wear rates of equiatomic TiNi and 60NiTi (aged, annealed and solution treated).

Material	Specific wear rate (mm^3/Nm)
Equiatomic TiNi	10.86×10^{-5}
Aged 60NiTi	7.02×10^{-5}
Annealed 60NiTi	9.73×10^{-5}
Solution treated 60NiTi	6.66×10^{-5}

Morphological analysis of equiatomic TiNi and 60NiTi worn surfaces and the tungsten carbide (WC) ball counter-face were investigated using scanning electron microscopy and electron dispersive spectroscopy to understand the operative wear mechanisms. Examination of the WC ball counter-face surface reveals clear evidence of adhesive wear. Adhesive wear occurs as a result of the formation of asperity junctions (cold weld) between the two surfaces in contact. This is followed by the shearing of the junction leaving material from one surface attached to the other surface. SEM micrograph and EDS results confirm the presence of TiNi layer on the WC ball surface as shown inFigure 8. Figure 8 insert shows Ti and Ni peaks appearing together with W, C, and Co.

SEM examination of aged 60NiTi wear tracks subjected to 60 N and 20 Hz revels evidence of plastic flow of material in the sliding direction, deep grooves oriented along the sliding direction and oxide formation on the wear tracks as shown in Figure 9. The presence of grooves and ridges on the wear track are evidence of abrasive wear. The grooves are a result of ploughing action by the counter-face and/or loose wear debris as third party abrasion. Abrasive wear seems to be more dominant in equiatomic TiNi, aged 60NiTi and solution treated 60NiTi. Besides adhesive and abrasive wear, evidence of delamination wear can also be observed from Figure 10. In delamination, wear takes place by heavy plastic deformation of the surface layers, nucleation of subsurface cracks where shear stress is maximum, and propagation of these cracks parallel to the surface. A surface pit forms as a result of delamination, as can be seen in Figure 9. Figure 10 clearly show the formation of plate-like wear particle by delamination. It is believed that delamination in superelastic alloys is promoted by the mismatch in the elastic properties across the sample cross section, which in turn, leads to accelerated delamination. During sliding wear, the top surface of the superelastic alloys experience heavy plastic deformation and loose their superelastic property, which causes martensite to stabilize; while, subsurface layers still behave in a superelastic manner. Hence, this mismatch in mechanical properties causes nucleation of cracks at the interface and propagation of these cracks parallel to the surface. Delamination wear seems to be more dominant in annealed 60NiTi. A high magnification image of the wear track in Figure 11 shows extensive Hertzian-type micro-cracks. Micro-cracks are developed due to surface tensile stress and extend perpendicular to the sliding direction. It was found that micro-cracks are more prominent at high load and high frequency.

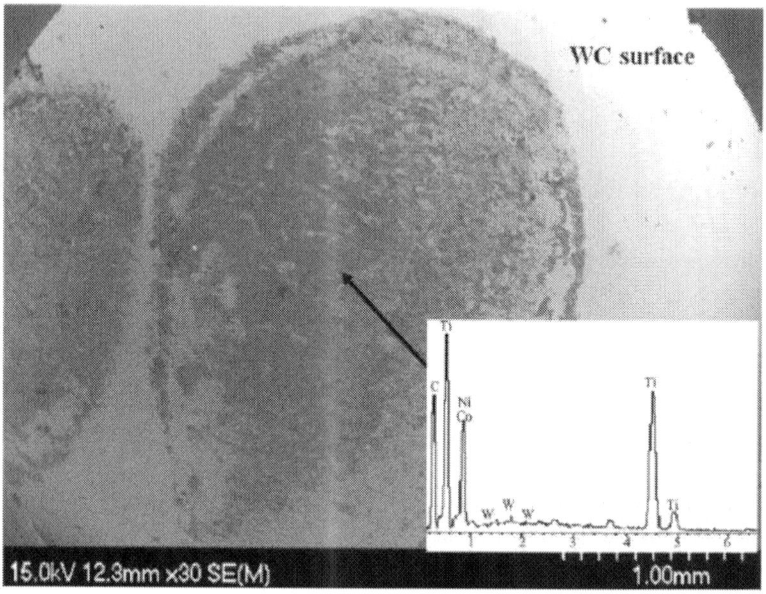

Figure 8. SEM micrograph of WC ball showing TiNi transfer layer.

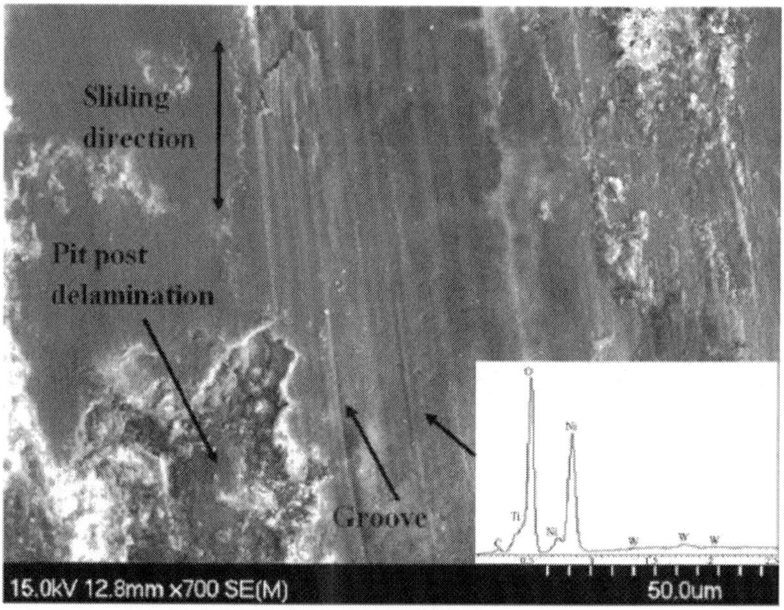

Figure 9. SEM micrograph of aged 60NiTi wear track subjected to 60 N load and 20 Hz frequency at the end of the wear test.

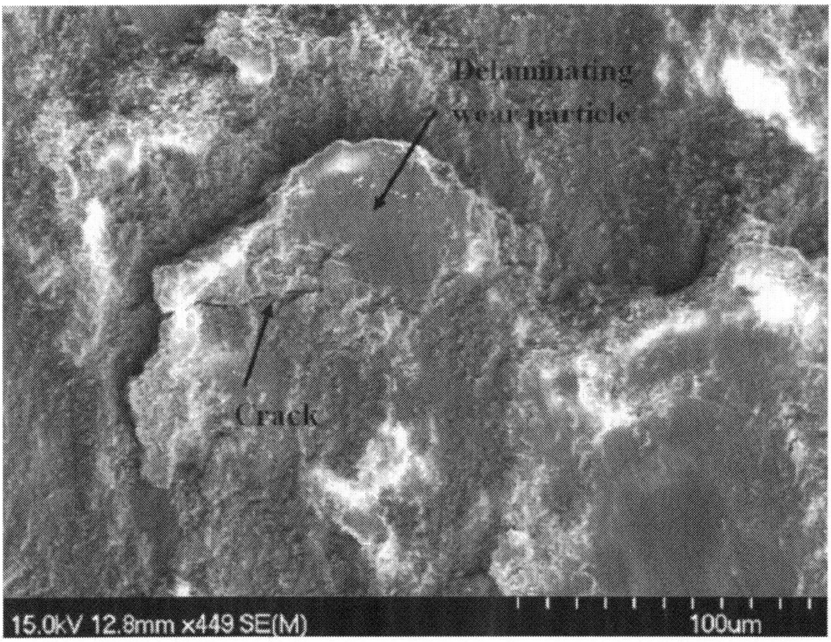

Figure 10. SEM image of aged 60NiTi wear track subjected to 60 N load and 20 Hz frequency at the end of the wear track with delaminating wear particle.

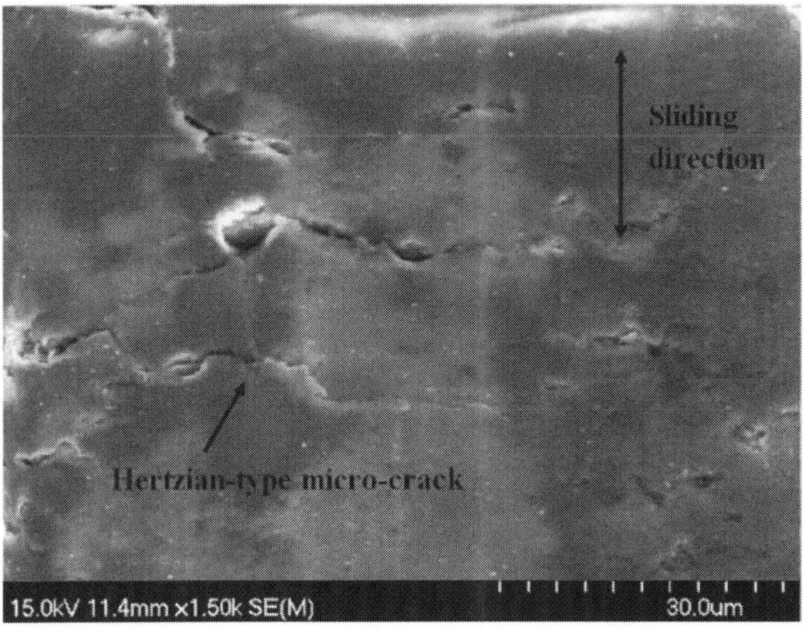

Figure 11. SEM images of annealed 60NiTi wear track subjected to 60 N load and 20 Hz frequency at the end of the wear test.

4. CONCLUSIONS

In the present work, dent and wear behaviors of equiatomic superelastic TiNi and 60NiTi are investigated and the following conclusions are reached:

1) Within the indentation condition employed in this study, the indentation behavior and dent resistance of TiNi and 60NiTi are not significantly affected by loading rate. Only a mild drop of about 5% - 10% in superelasticity is observed as the loading rate is increased. This drop may be due to retardation of martensitic transformation upon loading at high loading rates;
2) The superior wear and dent resistance of superelastic equiatomic TiNi and 60NiTi is attributed to low E/H and high elastic recovery ratio. The E/H value of TiNi is 15 while that of AISI 304 steel is 85. The average elastic recovery ratio for superelastic TiNi obtained from nanoindentation tests is over 7 times than that of AISI 304 stainless steel, although both have similar hardness;
3) Aged and solution treated new generation 60NiTi exhibits slightly enhanced superelasticity compared to equiatomic TiNi (i.e., lower E/H), yet they show about 3 times the hardness, slightly lower COF and about 40% improvement in wear resistance;
4) Adhesive, abrasive and delamination wear mechanisms have been identified. Adhesive wear and abrasive wear are more dominant in equiatomic TiNi, aged 60NiTi and solution treated 60NiTi, while adhesive wear and delamination wear are more dominant in annealed 60NiTi.

ACKNOWLEDGEMENTS

The authors would like to thank the Natural Sciences and Engineering Research Council of Canada (NSERC) for financial support, and Dr. Malcolm K. Stanford and Dr. Christopher DellaCorte, National Aeronautics and Space Administration (NASA), Glenn Research Center, Cleveland, Ohio for the preparation of the 60NiTi.

REFERENCES

1. Otsuka, K. and Ren, X. (2005) Physical Metallurgy of Ti-Ni Based Shape Memory Alloys. Progress in Materials Science, 50, 511-678.
2. Lagoudas, D., Ed. (2008) Shape Memory Alloys Modeling and Engineering Applications. Springer, New York.
3. Neupane, R. and Farhat, Z. (2013) Wear and Dent Resistance of Superelastic TiNi Alloy. Wear, 301, 682-687.

4. Li, D. (2000) Development of Novel Wear-Resistant Materials: TiNi-Based Pseudoelastic Tribomaterials. Materials & Design, 21, 551-555.
5. Li, D. (1998) A New Type of Wear-Resistant Material: Pseudo-Elastic TiNi Alloy. Wear, 221, 116-123.
6. Neupane, R. and Farhat, Z. (2014) Wear Mechanisms of Nitinol under Reciprocating Sliding Contact. Wear, 315, 25-30.
7. DellaCorte, C., Pepper, S., Noebe, R., Hull, D. and Glennon G. (2009) Intermetallic Nickel-Titanium Alloys for Oil-Lubricated Bearing Applications. Power Transmission Engineering, 8, 26-35.
8. Julien, G. (2005) Nitinol Ball Bearing Element and Process for Making. US Patent No. 6886986 B1.
9. Julien, G. (2006) Shape Memory Parts of 60 Nitinol. US Patent No. 7005018 B2.
10. Farhat, Z., Jarjoura, G., and Shahirnia, M. (2013) Dent Resistance and Effect of Indentation Loading Rate on Superelastic TiNi Alloy. Metallurgical and Materials Transactions A, 44, 3544-3551.
11. Farhat, Z. and Zhang, C. (2010) The Role of Reversible Martensite Transformation in the Wear Process of TiNi Shape Memory Alloy. Tribology Transactions, 53, 917-926.
12. Li, D. and Liu, R. (1999) The Mechanism Responsible for High Wear Resistancc of Pseudo-Elastic TiNi Alloy—A Novel Tribo-Material. Wear, 225-229, 777-783.
13. Lin, H., He, J., Chen, K., Liao, H. and Lin, K. (1997) Wear Characteristics of TiNi Shape Memory Alloys. Metallurgical and Materials Transactions A, 28, 1871-1877.
14. Archard, J.F. (1953) Contact and Rubbing of Flat Surfaces. Journal of Applied Physics, 24, 981-988.
15. DellaCorte, C., Moore III, L.E. and Clifton, J.S. (2013) The Effect of Pre-Stressing on the Static Indentation Load Capacity of the Superelastic 60NiTi. NASA TM 2013-216479.
16. Canter, N. (2014) Corrosion-Proof Nickel Titanium Bearings. Tribology & Lubrication Technology, 70, 10-11.
17. Wang, Z., Lei, H., Zhou, B., Wang, Y. and Zhang, C. (2011) Influence of Strain Rate on Mechanical Properties of Shape Memory Alloy. Key Engineering Materials, 467-469, 585-588.
18. Liu, Y., Li, Y. and Ramesh, K.T. (2002) Rate Dependence of Deformation Mechanisms in a Shape Memory Alloy. Philosophical Magazine A, 82, 2461-2473.
19. Vitiello, A., Giorleo, G. and Morace, R.E. (2005) Analysis of Thermomechanical Behaviour of Nitinol Wires with High Strain Rates. Smart Materials and Structures, 14, 215.

20. Lim, T.J. and McDowell, D.L. (1999) Mechanical Behavior of an Ni-Ti Shape Memory Alloy under Axial-Torsional Proportional and Nonproportional Loading. Journal of Engineering Materials and Technology, 121, 9-18.

21. Lin, P., Tobushi, H., Tanaka, K., Hattori, T. and Makita, M. (1994) Pseudoelastic Behaviour of TiNi Shape Memory Alloy Subjected to Strain Variations. Journal of Intelligent Material Systems and Structures, 5, 694-701.
22. Tobushi, H., Yoshirou, S., Takashi, H. and Kikuaki, T. (1998) Influence of Strain Rate on Superelastic Properties of TiNi Shape Memory Alloy. Mechanics of Materials, 30, 141-150.
23. Dayananda, G.N. and Rao, M.S. (2008) Effect of Strain Rate on Properties of Superelastic NiTi Thin Wires. Materials Science and Engineering: A, 486, 96-103.
24. Saletti, D., Pattofatto, S. and Zhao, H. (2010) Evolution of the Martensitic Transformation in Shape Memory Alloys under High Strain Rates. EPJ Web of Conferences, 14th International Conference on Experimental Mechanics, 6, Article No. 29008.
25. Shahirnia, M., Farhat, Z. and Jarjoura, G. (2011) Effects of Temperature and Loading Rate on the Deformation Characteristics of Superelastic TiNi Shape Memory Alloys under Localized Compressive Loads. Materials Science and Engineering: A, 530, 628-632.
26. Amini, A., He, Y. and Sun, Q. (2011) Loading rate Dependency of Maximum Nanoindentation Depth in Nano-Grained NiTi Shape Memory Alloy. Materials Letters, 65, 464-466.
27. Oliver, W. and Pharr, G. (1992) An Improved Technique for Determining Hardness and Elastic Modulus Using Load and Displacement Sensing Indentation Experiments. Journal of Materials Research, 7, 1564-1583.
28. ASTM Standard G133-05 (2010) Standard Test Method for Linearly Reciprocating Ball-on-Flat Sliding Wear. ASTM International, West Conshohocken.
29. Otsuka, K. and Shimizu, K. (1986) Pseudoelasticity and Shape Memory Effects in Alloys. International Metals Reviews, 31, 93-114.
30. Larsen-Basse, J. (1992) Introduction to Friction. In: Blau, P., Ed., ASM Handbook, Vol. 18: Friction, Lubrication, and Wear Technology, ASM International, Materials Park, 27-38.
31. Ashby, M.F. (2005) Materials Selection in Mechanical Design. Butterworth-Heinemann, Amsterdam.

32. Sen, S. and Sen, U. (2009) The Effect of Boronizing and Boro-Chromizing on Tribological Performance of AISI 52100 Bearing Steels. Industrial Lubrication and Tribology, 61, 146-153.

Chapter 5

Enhancing Mar and Abrasion Resistance of Acrylic Hard Coatings with Soft Base Layer

Patcharida Chouwatat1, Masaya Kotaki2,3, Atsushi Yokohama1*

[1]*Department of Advanced Fibro-Science, Kyoto Institute of Technology, Kyoto, Japan*
[2]*Kaneka US Material Research Center, Kaneka Americas Holding, Inc., Pasadena, Texas, USA*
[3]*Center for Fiber and Textile Science, Kyoto Institute of Technology, Kyoto, Japan*

ABSTRACT

Mar and abrasion resistance were investigated by a progressive load scratch test and steel wool abrasion test, respectively. Two acrylic coating systems including trimethylolpropane triacrylate (TMPTA) and pentaerythritol triacrylate (PETA) were prepared. A soft base layer was introduced as an intermediate layer between two different types of top layer and poly (methyl methacrylate) (PMMA) substrate to demonstrate the effect of soft base layer on mar and abrasion resistance. Abrasion damage on the coating surface was found to be less severe, when the soft base layer was incorporated into the coating systems. The reduction in scratch coefficient of friction (SCOF) and surface roughness was also observed. The results suggested that mar and abrasion resistance was greatly influenced by the presence of soft base layer, although different top layers were used. Moreover, it was found that abrasion resistance was further improved as the thicker soft base layer was applied.

Keywords: Acrylic Coating, Soft Base Layer, Mar Resistance, Abrasion Resistance

1. INTRODUCTION

Transparent polymeric materials have been utilized for various applications such as automobile windows, optical lenses and displays for electronic devices [1] - [5] . However, it is well-known that polymer materials having excellent optical transparency show poor abrasion resistance [6] -[8] . Thus, polymer hard coatings such as acrylic coatings have been developed to decorate and protect surface of polymer substrate without compromising transparency [9] - [11] . In addition, to optimize coating performance, a number of factors influencing abrasion resistance have been investigated, for example, coating compositions, coating and substrate characteristics, coating thickness, adhesion [4] [12] [13] , etc. It appears that those factors strongly influence abrasion resistance.

Therefore, a multilayer coating has been adopted to further improve the coating performance including chemical, mechanical and tribological properties [13] [14] . Each layer can provide a specific property, which has significant impact on tailoring the final coating performance. Moreover, it is found that combing coating layers with different mechanical properties can change the stress concentration in the surface area [13] . Sidorenko et al. demonstrated that a polymer trilayer film consisting of compliant rubber interlayer can be used as an abrasion resistant nanoscale coating with low friction coefficient for silicon surface [15] . This coating is prepared by a multiple grafting technique applied to self-assembled monolayers (SAM) and functionalized tri-block copolymer, followed by the photopolymerization of a topmost polymer layer. The compliant rubber interlayer can mediate localized stresses transferred through the top layer. Similarly, Vainshtein et al. emphasized an influence of soft base layer on scratch and abrasion resistance for oxide coating [16] . Polydimethylsiloxane (PDMS) is applied as soft intermediate layer between polycarbonate (PC) and titania. The presence of PDMS soft layer resulted in enhanced abrasion resistance and reduction in friction coefficient due to cushioning effect, which reduces the local stress. These studies indicate that soft base layer greatly influences abrasion resistance in nanoscale. However, the effect of soft base layer on macro-abrasion resistance and mar behavior for polymeric coating on polymer substrate has not been clarified yet.

Hence, herein, acrylic coatings, i.e. trimethylolpropane triacrylate (TMPTA) and pentaerythritol triacrylate (PETA), with and without soft base layer are applied on poly (methyl methacrylate) (PMMA) substrate. The influence of soft base layer on mar and abrasion behavior of acrylic coatings is investigated by using a progressive load scratch test (ISO 19252) and steel wool abrasion test, respectively. Moreover, the effect of soft base layer thickness on abrasion resistance is further demonstrated.

2. EXPERIMENTAL

2.1. Coating Preparation

Single- and bilayer coatings were prepared using UV curing technique. The formulations of each coating are listed in Table 1. A solution of top layer, i.e. trimethylolpropane triacrylate (TMPTA) and pentaerythritol triacrylate (PETA) was mixed with 5 wt% of photoinitiator, and then coated on 2 mm-thick poly (methyl methacrylate) (PMMA) substrates using Mayer rod coating. The wet film was cured under ultraviolet light to obtain TMPTA and PETA single-layer coatings (TS and PS), respectively. The thickness of top layer was set at 30 μm.

For bilayer coatings, a soft base layer was introduced to the coating systems. A solution of soft base layer with 5 wt% of photoinitiator was applied on PMMA substrate, and exposed under ultraviolet light. The thickness of soft base layer was 5 and 15 μm. The solution of the top layer with 5% of photoinitiation was then coated on the soft base layer, and cured under the UV light to obtain the bilayer coatings with different soft base layer thickness: TB5, TB15, PB5, PB15.

Table 1. Formulations and critical normal load of scratch visibility for acrylic coatings.

Sample ID	Coating formulations		Critical normal load for scratch visibility (N)
	Top layer materials	Soft base layer thickness (μm)	
TS		-	5.8
TB5	TMPTA	5	Over 20 N
TB15		15	Over 20 N
PS		-	Over 20 N
PB5	PETA	5	Over 20 N
PB15		15	Over 20 N

2.2. Characterization

Surface hardness of each layer was measured by direct imaging indenter (DII, SANKO Co., LTD) combining with a video camera and a transparent Berkovich tip to directly measure contact area. The soft base layer, TMPTA and PETA, was coated on glass substrates. Coating layer thickness was set at 25 μm. The test was conducted at room temperature under a load control mode set in the range from 0 to 30 mN (5 sec) and from 30 to 0 mN (5 sec). The hardness was calculated by the ratio of load to indented area.

Tensile properties of each layer were measured using a Universal Tensile Testing Machine (Instron model 4206). Free-standing films were prepared by UV curing technique. The dimension of specimens was 50 mm by 15 mm. The thickness of specimen was 0.2 mm. The test was conducted at room temperature with crosshead speed of 0.085 mm/s and a gauge length of 15 mm.

The progressive load scratch test according to ISO 19252 was conducted using a scratch instrument model KK-02 (KATO Tech Co.) at room temperature. Normal load was linearly increased from 1 to 20 N. Scratch velocity and scratch length were set at 100 mm/sec and 70 mm, respectively. A 10 mm diameter of stainless steel ball tip was used to study on soft base layer effect on mar resistance. Five tests were performed, and the values obtained were averaged for each sample.

In addition, steel wool abrasion test was carried out at room temperature using a scratch instrument model KK-01 (KATO Tech Co.). The coating surface was rubbed for 100 cycles with #0000 steel wool under constant load of 5 N and scratch velocity of 100 mm/sec. Scratch length was set at 50 mm.

To evaluate abrasion resistance after scratch and abrasion test, haze was measured with NDH5000 haze meter (Nippon Denshoku Industries Co. Ltd.).

VK-X100 laser microscope (Keyence Corporation) was used to observe damage on the coating surface and surface roughness. The multi-file analysis software provided with the microscope was used to measure the surface roughness inside the abrasion path. The tilt correction available in the software was processed before measuring the surface roughness.

The more detailed damage was analyzed using field emission scanning electron microscopy (FE-SEM, Hitachi S4200) at an accelerating voltage at 8 kV. The coating samples were cut around the interesting point of damage, dried in vacuum overnight and coated with gold.

3. RESULTS AND DISCUSSION

3.1. Mar Behavior

Acrylic coatings with and without soft base layer were prepared to clarify the effect of soft base layer on mar and abrasion resistance. Hardness and tensile properties of each coating layer were evaluated. The results are listed in Table 2. Hardness of soft base layer was about 50% lower than that of TMPTA and PETA whereas in top layer TMPTA had 9% higher hardness than PETA. In addition, the soft base layer showed lowest tensile modulus and largest elongation at break, while tensile strength at break of soft base layer was slightly higher than that of TMPTA but lower than that of PETA.

To study the effect of soft base layer on mar resistance, the progressive load scratch test according to ISO 19,252 was performed with 10 mm in diameter stainless steel ball tip. Scratched surfaces of each coating were carefully observed using laser microscope and SEM. The detailed damage at different Normal load for single- and bilayer coatings is showed in Figure 1 and Figure 2, respectively. Single layer coatings, i.e. TS and PS, started to have damages at very low Normal load and the damages became more severe with increasing Normal load (Figure 1). Obviously, in bilayer coatings (TB15 and PB15), mar damages were Not seen at low Normal load but became visible when normal load further increased up to 20 N (Figure 2). In addition, the effect of introduction of soft base layer on mar resistance is readily found in Figure 3. All bilayer coatings showed less severe mar damages than TS and PS. Moreover, the thicker soft base layer was introduced, the milder mar damages were observed. These indicate that the introduction of soft base layer significantly affects mar behavior, although the top layer is different.

Table 2. Surface hardness and tensile properties of each coating layer.

Sample ID	Surface hardness (MPa)	Tensile properties		
		Tensile modulus (GPa)	Tensile strength at break (MPa)	Elongation at break (%)
TMPTA	267	2.3	51.1	3.1
PETA	242	2.9	64.4	3.3
Soft base layer	132	1.5	53.9	6.0

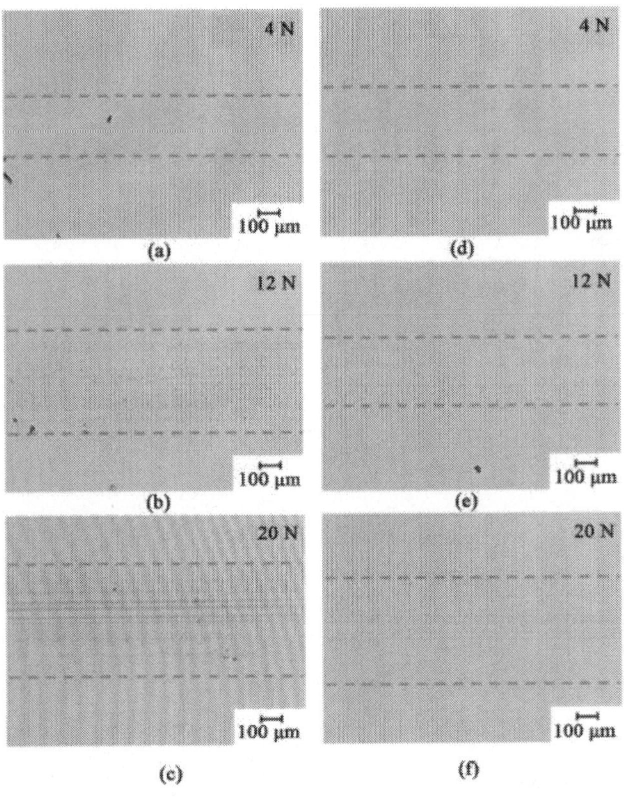

Figure 1. Mar damage on single-layer coating surfaces: (a)-(c) TMPTA and (d)-(f) PETA coating system.

The above results are greatly correlated to the scratch coefficient of friction (SCOF) curves in Figure 4. The SCOF curves were calculated by the ratio of the tangential force to the normal force [17] . It clearly sees that in both coating systems, the bilayer coatings had lower SCOF than the single layer coatings. This might be due to shallower penetration of scratch tip and less severe damage on the coating surface. For TMPTA system, the thicker soft layer resulted in reduction of SCOF. However, it was found that PB15 had higher SCOF than PB5 but showed milder damage (Figure 3). This might be due to wear debris that transfers from the coating surface to scratch tip and make the tip move more smoothly.

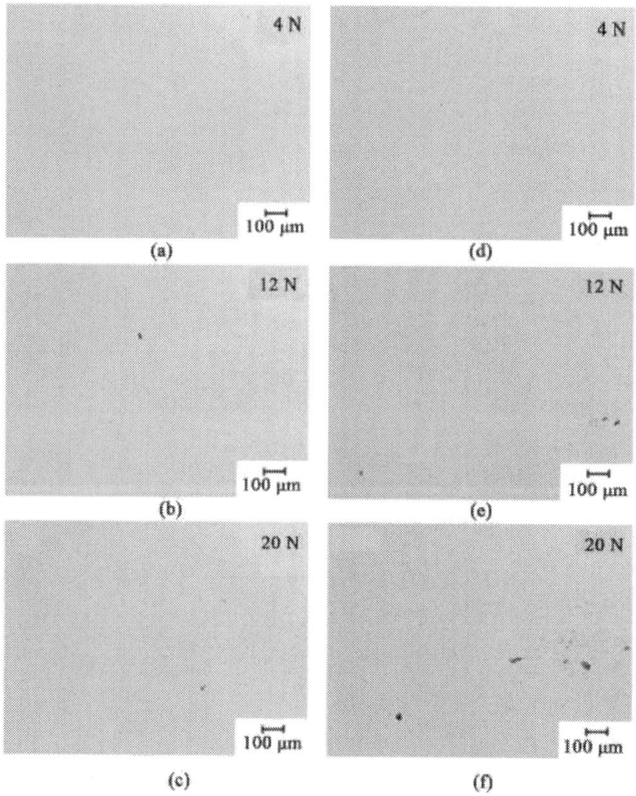

Figure 2. Mar damage on bi-layer coating surfaces: (a)-(c) TMPTA and (d)-(f) PETA coating system.

Since the surface damage may induce a change of the surface roughness that results in mar visibility and loss of gloss and/or transparency [13] [18] , an onset of mar visibility was determined by naked eyes and then converted to a critical normal load for onset of mar visibility as presented inTable 1. A great improvement in mar visibility resistance was found as indicated by an increase in critical Normal load for onset of mar visibility when the soft layer was introduced into TMPTA coating system, while all coatings in PETA system exhibited excellent resistance to mar visibility.

However, the observation of mar visibility by naked eyes depends on many factors such as observation environmental condition and inspectors' sensitivity, making the evaluation more complicated and inconsistent [19] . Thus, to confirm the above result, haze measurement was performed. Figure 5 illustrates the relationship between haze values of each coating and normal load. It was found that the trends of haze values related well to the mar visibility result. For TMPTA system, the haze values in TS increased dramatically as normal load increased whereas TB5, TB15 and PETA system showed similar slopes of haze values and a gradual increase in haze values, relating to better mar visibility and mar resistance.

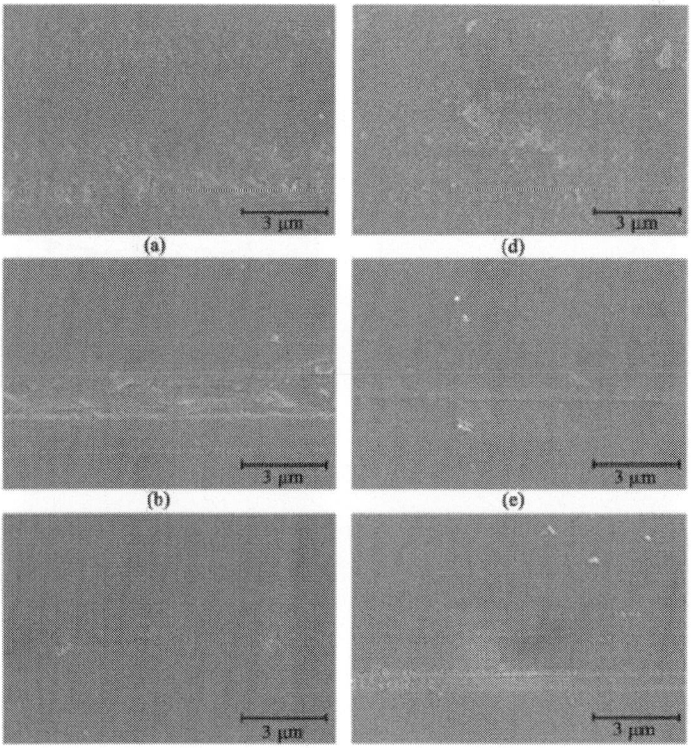

(a) (d)

(b) (e)

(c) (f)

Figure 3. High-magnification SEM micrographs showing mar damage on coating surfaces at 20 N: (a) TS; (b) TB5; (c) TB15; (d) PS; (e) PB5 and (f) PB15.

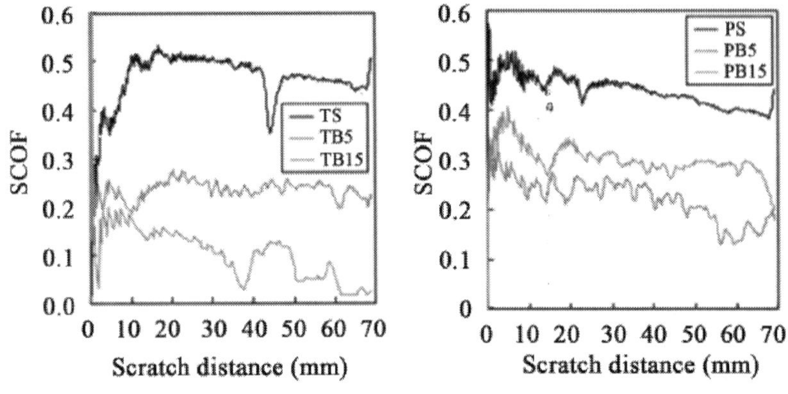

Figure 4. Comparison of SCOF curves from progressive load scratch test: (a) TMPTA and (b) PETA coating system.

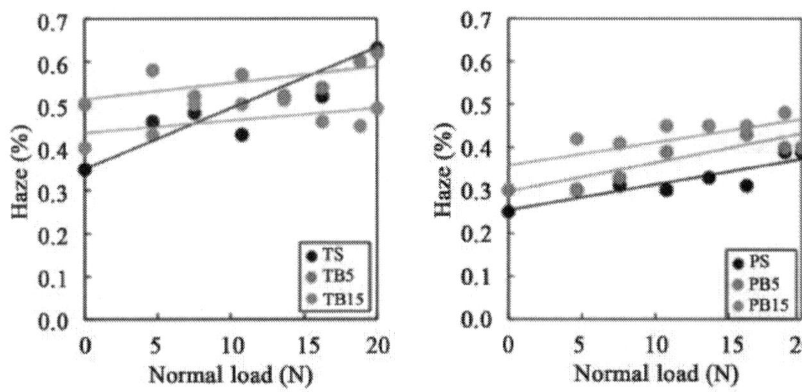

Figure 5. Haze values measured after progressive load scratch test: (a) TMPTA and (b) PETA coating system.

To clarify mechanism of mar behavior, surface roughness (S_a) on the scratch groove at 20 N of normal load was measured. Figure 6 compares the surface roughness of each coating. Similar trend was observed in both coating systems. The introduction of soft base layer caused a decrease in surface roughness. Moreover, the surface roughness decreased with increasing the soft base layer thickness. It was also found that the surface roughness of TMPTA coating system was much higher than those of PETA coating system. This might be explained by the tensile properties. As the tensile properties play an important role in delaying or preventing an increase of scratch-induced surface roughness, the higher tensile strength leads to lower surface roughness on the scratch groove [13] [19] [20] . Correspondingly, PETA layer has higher modulus and tensile strength than TMPTA. Thus, PETA coating system shows lower surface roughness, correlating strongly with better resistance to mar visibility and mar.

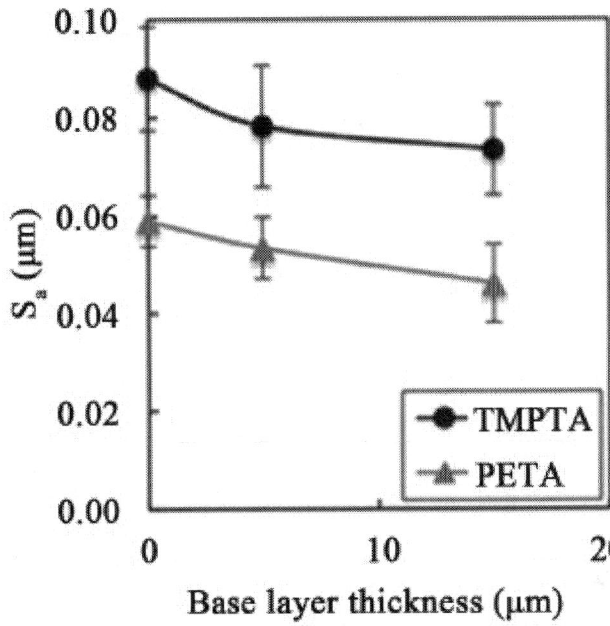

Figure 6. Surface roughness measured after progressive load scratch test.

3.2. Abrasion Behavior

A steel wool abrasion test has been widely used in industries to evaluate the coating performance. In this work, a steel wool abrasion test was also utilized to verify the soft base layer effect. The relationship between the SCOF obtained from steel wool abrasion test and Number of test cycles is shown in Figure 7. The SCOF curves shot up at the beginning of the test and steady increased with further increasing test cycles. In TMPTA system, at the early abrasion test TS showed lowest SCOF whereas TB5 had highest SCOF (Figure 7(a)). The SCOF curves started to converge after 20 cycles. Conversely, PETA coating system exhibited the similar trend of SCOF at the early abrasion test but SCOF of PS diverged and was lower than PB5 and PB15 after 20 cycles (Figure 7(b)). This might be due to the fact that when the Number of test cycles increased, the large amount of debris transferred from the coating surface to steel wool, making the SCOF relatively low.

Figure 8 presents the haze values after steel wool abrasion test. It was found that the haze values of TS increase dramatically during the test while the haze values of TB5 and TB15 steadily increased after 30 cycles (Figure 8(a)). Moreover, considering the haze values after 100 cycles, it was found that the introduction of soft base layer into TMPTA system resulted in a significant decrease in haze values, relating to an improvement of abrasion resistance. The thicker soft layer led to better performance. However, the haze values in PETA coating system exhibited similar trend: they linearly increased as the number of cycles increased (Figure 8(b)).

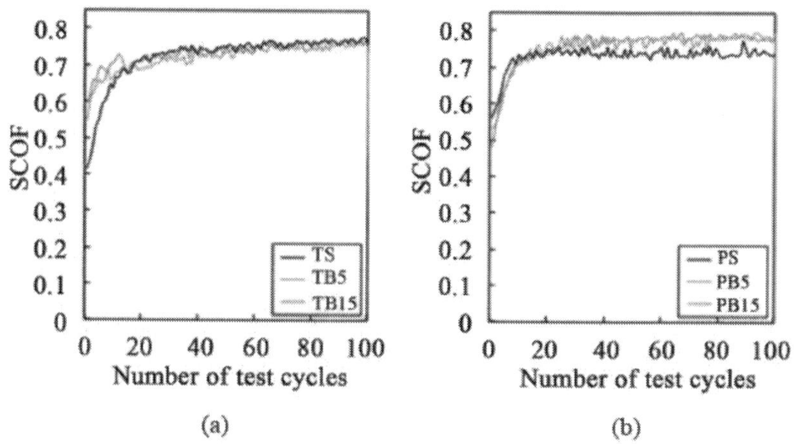

Figure 7. Comparison of SCOF curves from steel wool abrasion test: (a) TMPTA and (b) PETA coating system.

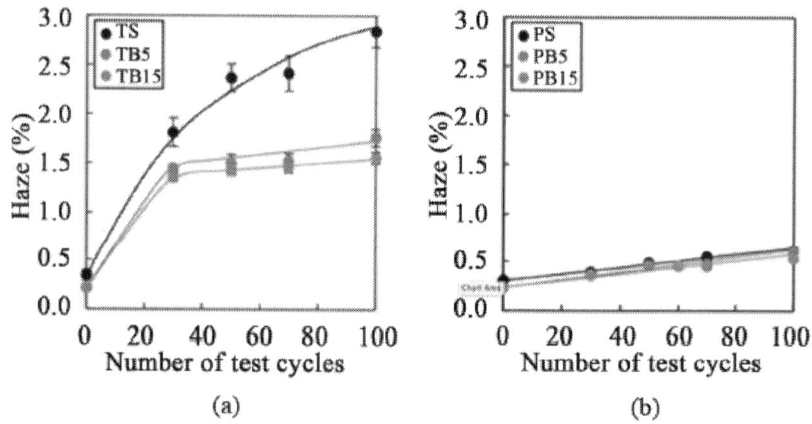

Figure 8. Haze values as a function of number of abrasion cycles: (a) TMPTA and (b) PETA coating system.

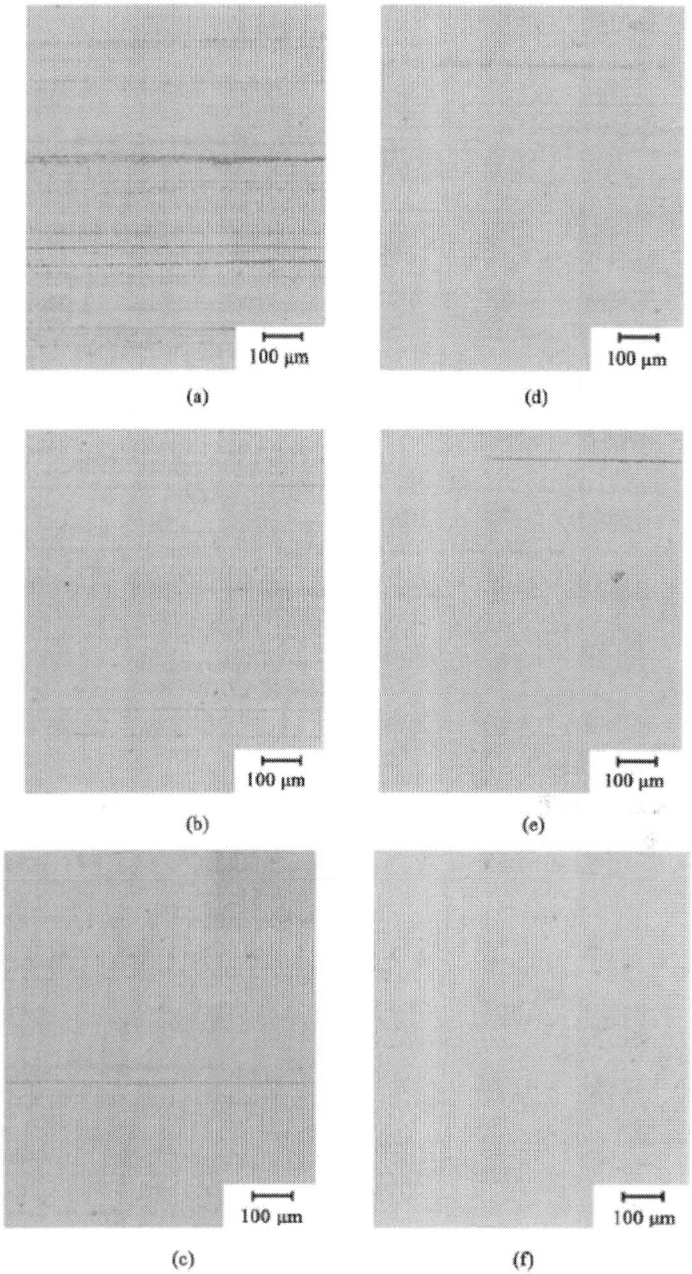

Figure 9. Abrasion damage observed after steel wool abrasion test: (a) TS; (b) TB5; (c) TB15; (d) PS; (e) PB5 and (f) PB15.

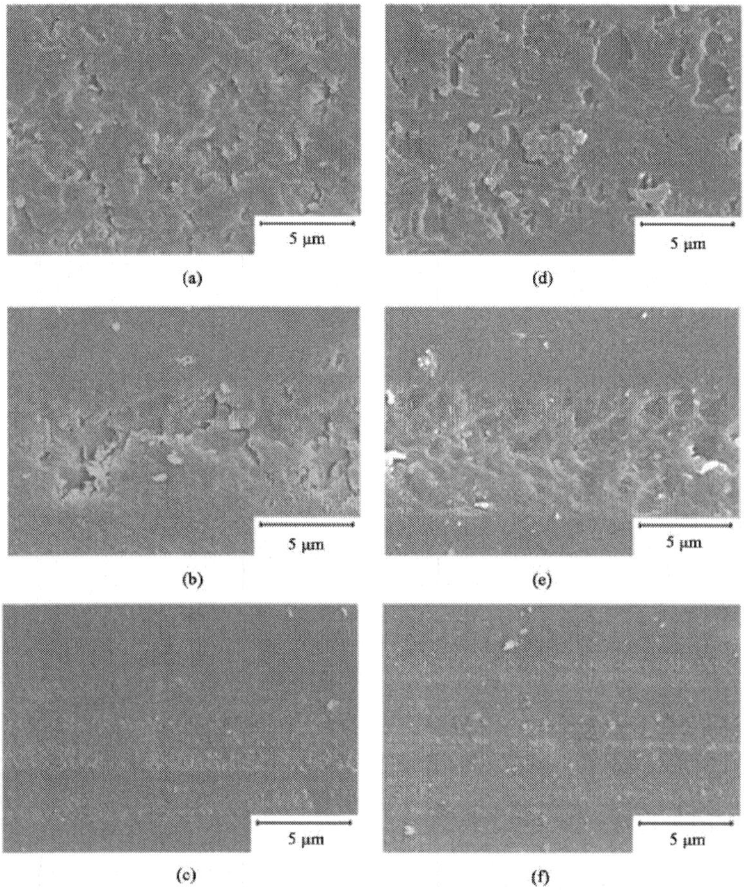

Figure 10. SEM micrographs showing abrasion damage from steel wool abrasion test: (a) TS; (b) TB5; (c) TB15; (d) PS; (e) PB5 and (f) PB15.

The effect of introduction of soft base layer on abrasion behavior is readily observed in Figure 9 andFigure 10. Bilayer coatings showed less severe damage than single layer coatings. Moreover, the introduction of 15 μm-thick soft base layer into both coating systems led to only minor abrasion damage (Figure 10(c) and Figure 10(f)). It is clearly evident that the presence of soft base layer in both coating systems causes marked decline in the surface roughness after 100 cycles (Figure 11), relating to enhanced abrasion resistance.

This present study emphasizes that abrasion resistance is strongly influenced by the soft base layer despite using different top layers and test methods. The thicker soft base layer results in better abrasion resistance. Furthermore, the SCOF decreases when the soft base layer is introduced. This might suggest that the introduction of soft base layer can minimize the localized stress at the top surface due to the stress distribution. The thicker soft base layer results in less localized stress concentrated at the top surface.

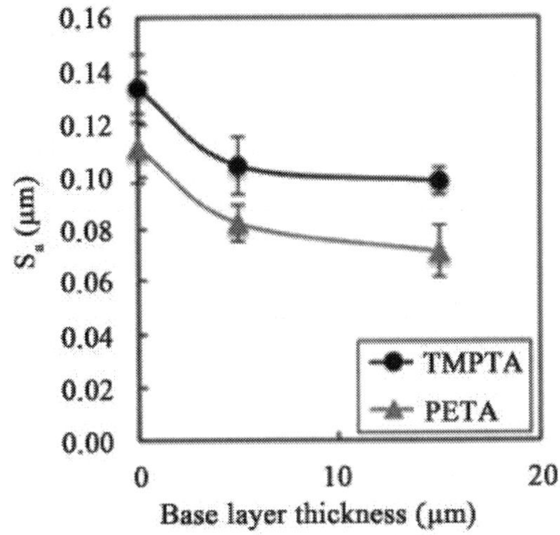

Figure 11. Surface roughness measured after steel wool abrasion test.

4. CONCLUSION

The soft base layer effect on mar and abrasion resistance of acrylic coatings was investigated using progressive load scratch test and steel wool abrasion test. Both test method indicated that mar and abrasion resistance was greatly influenced by the presence of soft base layer. The SCOF and surface roughness significantly reduced when the coatings consisted of soft base layer. The introduction of soft base layer resulted in a marked improvement in mar and abrasion resistance, despite the fact that top layers were different. Furthermore, it was found that the thickness of soft base layer also affected mar and abrasion resistance. The thicker the soft base layer is, the better the abrasion resistance can be observed.

ACKNOWLEDGEMENTS

The authors would like to acknowledge Japan Scratch Consortium Society for insightful discussion regarding this research. We thank SANKO Co., Ltd. for conducting indentation test. We also thank Nippon Denshoku Industries Co. Ltd. for their support of haze meter in this study.

REFERENCES

1. Schottner, G., Rose, K. and Posset, U. (2003) Scratch and Abrasion Resistant Coatings on Plastic Lenses—State of the Art, Current Developments and Perspectives. Journal of Sol-Gel Science and Technology, 27, 71-79.
2. Schulz, U., Volker, W., Thomas, K. and Paul, A. (2001) The Influence of Weathering on Scratches and on Scratch and Mar Resistance of Automotive Coatings. Progress in Organic Coatings, 42, 38-48.
3. Hozumi, A., Kato, Y. and Takai, O. (1996) Two-Layer Hard Coatings on Transparent Resin Substrates for Improvement of Abrasion Resistance. Surface and Coatings Technology, 82, 16-22.
4. Mackenzie, J.D. and Bescher, E. (2003) Some Factors Governing the Coating of Organic Polymers by Sol-Gel Derived Hybrid Materials. Journal of Sol-Gel Science and Technology, 27, 7-14.
5. Rao, K.N. (2003) Studies on Thin Film Materials on Acrylics for Optical Applications. Bulletin of Materials Science, 26, 239-245.
6. Wouters, M.E.L., Wolfs, D.P., van der Linde, M.C., Hovens, J.H.P. and Tinnemans, A.H.A. (2004) Transparent UV Curable Antistatic Hybrid Coatings on Polycarbonate Prepared by the Sol-Gel Method. Progress in Organic Coatings, 51, 312-319.
7. Kim, S.W. (2010) Characterization of UV Curable Hybrid Hard Coating Materials Prepared by Sol-Gel Method. Korean Journal of Chemical Engineering, 28, 298-303.
8. Gilberts, J., Tinnemans, A.H.A., Hogerheide, M.P. and Koster, T.P.M. (1998) UV Curable Hard Transparent Hybrid Coating Materials on Polycarbonate Prepared by the Sol-Gel Method. Journal of Sol-Gel Science and Technology, 11, 153-159.
9. Schroeter, S.H. and Daniel, R.O. (1980) Abrasion Resistant Silicone Coated Polycarbonate Article. US Patent No. 4210699.
10. Ogawa, M. and Hara, Y. (1988) Silicone-Based Coating Composition for Surface-Releasing Film. US Patent No. 4764576.
11. Azuma, I., Kosaka, N., Iwamura, G., Marutani, Y. and Uemura, H. (1997) Acrylic Oligomer for High Solid Automotive Top Coating System Having Excellent Acid Resistance. Progress in Organic Coatings, 32, 1-7.
12. Wu, L., Guo, X. and Zhang, J. (2014) Abrasive Resistant Coatings—A Review. Lubricants, 2, 66-89.
13. Friedrich, K. and Schlarb, A.K. (2013) Tribology of Polymeric Nanocomposites: Friction and Wear of Bulk Materials and Coatings. Butterworth-Heinemann, Oxford.

14. Holmberg, K., Matthews, A. and Ronkainen, H. (1998) Coatings Tribology—Contact Mechanisms and Surface Design. Tribology International, 31, 107-120.
15. Sidorenko, A., Ahn, H.-S., Kim, D.-I., Yang, H. and Tsukruk, V.V. (2002) Wear Stability of Polymer Nanocomposite Coatings with Trilayer Architecture. Wear, 252, 946-955.

16. Gotlib-Vainshtein, K., Girshevitz, O., Sukenik, C.N., Barlam, D. and Cohen, S.R. (2014) A Nanometric Cushion for Enhancing Scratch and Wear Resistance of Hard Films. Beilstein Journal of Nanotechnology, 5, 1005-1015.
17. Browning, R., Lim, G.T., Moyse, A., Sun, L. and Sue, H.-J. (2006) Effects of Slip Agent and Talc Surface-Treatment on the Scratch Behavior of Thermoplastic Olefins. Polymer Engineering & Science, 46, 601-608.
18. L&oumlrinczová, I. and Decker, C. (2014) Scratch Resistance of UV-Cured Acrylic Clearcoats. Surface Coatings International Part B: Coatings Transactions, 89, 133-143.
19. Jiang, H., Browning, R.L., Hossain, M.M., Sue, H.-J. and Fujiwara, M. (2010) Quantitative Evaluation of Scratch Visibility Resistance of Polymers. Applied Surface Science, 256, 6324-6329.
20. Hossain, M., Browning, M., Minkwitz, R. and Sue, H.-J. (2012) Effect of Asymmetric Constitutive Behavior on Scratch-Induced Deformation of Polymers. Tribology Letters, 47, 113-122.

Chapter 6

Study of Radio Frequency Plasma Treatment of PVDF Film Using Ar, O_2 and (Ar + O_2) Gases for Improved Polypyrrole Adhesion

Akif Kaynak[1,], Tariq Mehmood[1], Xiujuan J. Dai[2], Kevin Magniez[2] and Abbas Kouzani[1]*

[1] *School of Engineering, Deakin University, Geelong, Victoria 3216, Australia*
[2] *Institute for Frontier Materials, Deakin University, Geelong, Victoria 3216, Australia*

ABSTRACT

Improvement of the binding of polypyrrole with PVDF (polyvinylidene fluoride) thin film using low pressure plasma was studied. The effects of various plasma gases *i.e.*, Ar, O_2 and Ar + O_2 gases on surface roughness, surface chemistry and hydrophilicity were noted. The topographical change of the PVDF film was observed by means of scanning electron microscopy and chemical changes by X-ray photoelectron spectroscopy, with adhesion of polypyrrole (PPy) by abrasion tests and sheet resistance measurements. Results showed that the increase in roughness and surface functionalization by oxygen functional groups contributed to improved adhesion and Ar + O_2 plasma gave better adhesion.

Keywords: coatings; plasma treatment; surface modification; polypyrrole; interfacial bonding; conducting polymers; PVDF

1. INTRODUCTION

PVDF (polyvinylidene fluoride) is a semi-crystalline polymer in which each monomer has two dipole moments, one due to CF_2 and the other due to CH_2. The most common and most studied crystal structures are α and β phases [1,2]. The piezo-electric and pyro-electric properties of PVDF mainly depend on the β-phase content and its growth can be induced by several techniques, the most common being the mechanical stretching of the α-phase films at a suitable temperature [3]. Mechanical strains due to the bending, vibration or compression of the thin-film structure can be the source of the energy generation [4,5]. A PVDF sensor was integrated into an automotive seat cover fabric to obtain information about the car occupant [1]. Hutchison *et al.* [6] used platinum coating in between PVDF and polypyrrole (PPy) to get a stable connection in their actuator device. Choi *et al.* [7] used conductive fabric and PVDF to monitor cardiorespiratory signals employing the piezoelectric property of PVDF. The construction of a PVDF based energy harvesting device require electrodes on both sides of the PVDF film, these electrodes are connected to an electronic circuit to harvest electricity during motion. Different metal coatings can be used as electrode materials [8] and are brittle and uncomfortable to the skin. PPy coatings can be used instead of metals for this purpose due to ease of coating, comfort, and conformity [9]. Bhat *et al.* [10] synthesized a PVDF PPy composite film by electrochemical polymerization and Mansouri *et al.* [11] coated PPy onto PVDF microfiltration membranes by an *in situ* chemical polymerization method.

One of the perceived problems for PPy coating is poor adhesion due to low surface energy and the chemical nature of PVDF. Plasma can be used for the surface modification to improve subsequent coating deposition and bonding [12,13]. Exposing PVDF films to argon (Ar), hydrogen (H_2), and oxygen (O_2) remote plasma, introduced various functional groups [14]. Substitution of carbonyl or hydroxyl groups with fluorine atoms (de-fluorination) is central to the improvement of adhesion [14,15,16,17]. Duca *et al.* [17] investigated radio frequency (RF) Ar plasma treatment on PVDF. The water contact angle and fluorine to carbon (F/C) ratio was decreased, while surface roughness increased after the argon plasma treatment. Ar and O_2 plasma treatments [18] incorporated oxygen onto the surfaces of PVDF. Our previous investigation [19] on a PET surface suggests that the improvement of adhesion of PPy was mainly due to incorporation of oxygen based functional groups on the surface after plasma treatment.

In this work, results of our investigations on PVDF under RF plasma for the improvement of adhesion of PPy are reported using Ar, O_2 and (Ar + O_2) gases. The surface morphology, wettability, chemical composition, electrical resistivity and abrasion resistance of PVDF were studied before and after each plasma treatment.

2. EXPERIMENTAL PROCEDURE

2.1. Materials

The substrate was a PVDF thin film of 50 micron thickness. Films were immersed in acetone, ultrasonicated for 5 min, then washed with deionized water and dried at room temperature. Substrate materials were then conditioned under standard atmospheric pressure at 65% ± 2% relative humidity and 21 ± 1 °C for at least 24 h prior to further processing. Monomer (pyrrole), oxidant (ferric chloride $FeCl_3 \cdot 6H_2O$) and dopant (para-toluene sulphonic acid mono hydrate, pTSA) reagent quality were purchased (Sigma-Aldrich, Melbourne, Australia).

2.2. Plasma Treatment

Plasma treatment of both PVDF thin films was performed in a custom built 13.5 MHz inductively coupled reactor [20]. The base pressure 1×10^{-3} mbar was achieved by use of a rotary pump. Ar plasma was used for the pre-treatment of the samples (100 W, 8×10^{-2} mbar, 30 s) to activate and clean the surface. Plasma power was kept at 100 W for all the gases. The gas pressure was set at8 $\times 10^{-2}$ mbar for Ar and 8×10^{-2} mbar for both O_2 and Ar + O_2 samples. In case of Ar + O_2, the composition of both gases was kept at 1:1 by volume. The plasma treatment time was fixed at 120 s for all the samples.

2.3. Contact Angle Measurement

Water contact angle (WCA) measurements were performed using a CAM100 WCA tester (KSV Instruments, Helsinki, Finland) and distilled water. The films were mounted on glass slides prior to measurement using double side adhesive tape. At least six measurements of each PVDF sample condition were taken to get an averaged WCA measurement. Attention Theta v4.1.9.8 (Biolin scientific, Stockholm, Sweden) software was used to calculate WCA from the captured images after 0.80 s of water droplet contact time. All the measurements were carried out within 10 min of plasma treatment.

2.4. Scanning Electron Microscopy

A Supra 55 VP (Carl Zeiss, Oberkochen, Germany) scanning electron microscope (SEM) was used to study the surface morphology of the PVDF thin film surface before and after plasma treatment. All the samples were gold coated before viewing in the SEM.

2.5. X-ray Photo Spectroscopy (XPS)

Spectra were obtained using a K-α X-ray photoelectron spectrometer (Thermo Fischer Scientific, Waltham, MA, USA) with monochromatic X-rays focused to a 400 μm spot size. Excessive charging of the samples was minimised using a flood gun. The binding energies of the samples were accurately established by charge shift; the binding energy peak of the F1s to 688.65 eV [14]. Survey spectra were obtained at pass energy of 100 eV while high resolution peak scans were performed at 20 eV pass energy. X-ray photoelectron spectroscopy (XPS) measurements were done within one week of plasma treatment.

2.6. Abrasion Resistance and Sheet Resistance

Circular PPy coated PVDF thin film samples were cut using a GSM (grams per square meter) cutter with 38 mm diameter and were abraded against a reference wool fabric under 9 kPa head weight in a Martindale abrasion tester (I.D.M instruments, Melbourne, Australia) for 0, 200 and 2000 abrasion cycles. Micrographs of the samples before and after abrasion were taken under the same lighting and exposure by a microscope (Olympus DP-71, Tokyo, Japan). Sheet resistance was measured using a SP4 four point probe head (Signatone, Tomkins Court Gilroy, CA, USA) and source meter (Keithley 2400, Solon, OH, USA) and employing the same pressure for all the samples [21]. Sheet resistance values were obtained from six different locations of each sample and later average values were calculated.

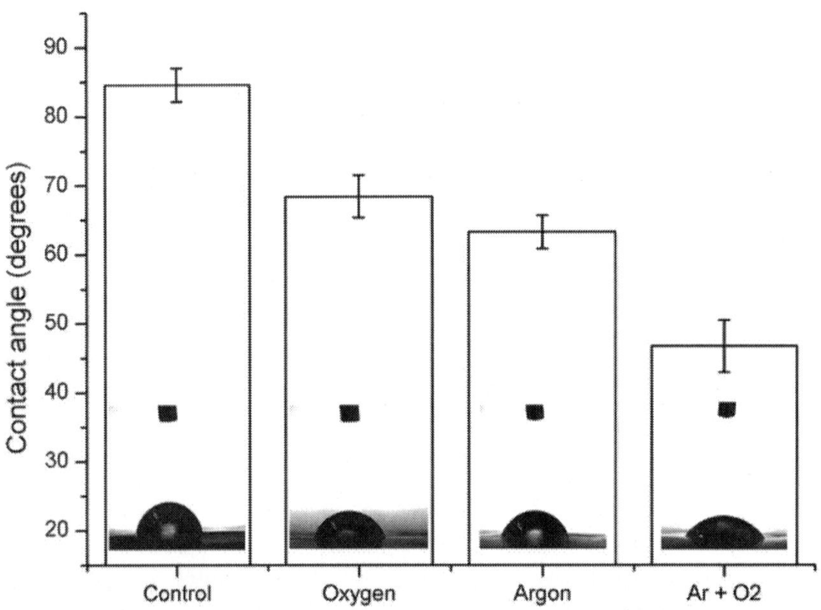

Figure 1. Water contact angle (WCA) of polyvinylidene fluoride (PVDF) thin film before and after Ar, O_2 and (Ar + O_2) plasma treatment.

3. RESULTS AND DISCUSSION

3.1. Contact Angle Measurement

Contact angle measurements show an improvement in the PVDF surface wettability after the plasma treatment. The average WCA value of the PVDF control sample was 84.55° and it reduced with plasma treatment irrespective of the gas used .WCA reduced to 68.36, 63.2 and 46.65 for Ar, O_2 and Ar + O_2 plasma samples respectively as shown in Figure 1. WCA is directly related to the surface energy, it is also dependent on surface morphology and chemical functional groups presentk on the surface.

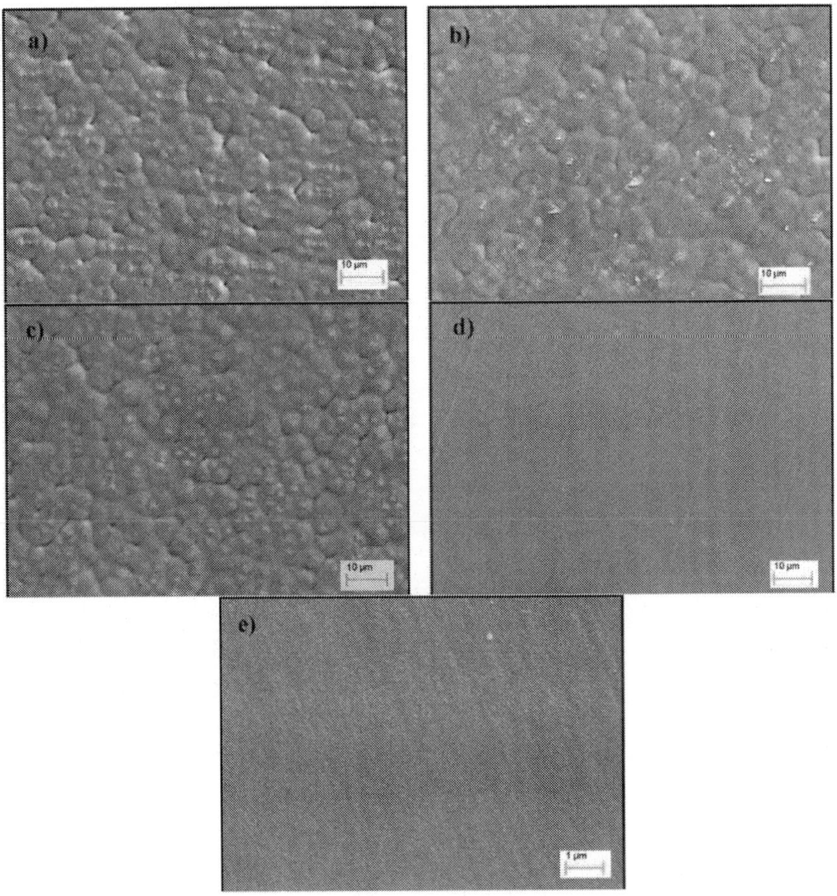

Figure 2. Scanning electron microscope (SEM) micrograph of polyvinylidene fluoride (PVDF): (**a**) control; (**b**) Ar plasma; (**c**) O_2 plasma and (**d,e**) (Ar + O_2) plasma.

3.2. Scanning Electron Microscopy

Changes in the surface morphology are normally caused by ion bombardment and UV radiation in the plasma and depend on plasma gas. The SEM images in Figure 2a–e show the surface of the PVDF film before and after plasma treatment. The grainy features on the PVDF film surface were carried onward from the film manufacturing process.

There is no visible etching by Ar or O_2 plasma whereas (Ar + O_2) plasma treatment significantly altered the surface morphology (Figure 2d). A rougher surface was displayed as shown in higher resolution image (Figure 2e).

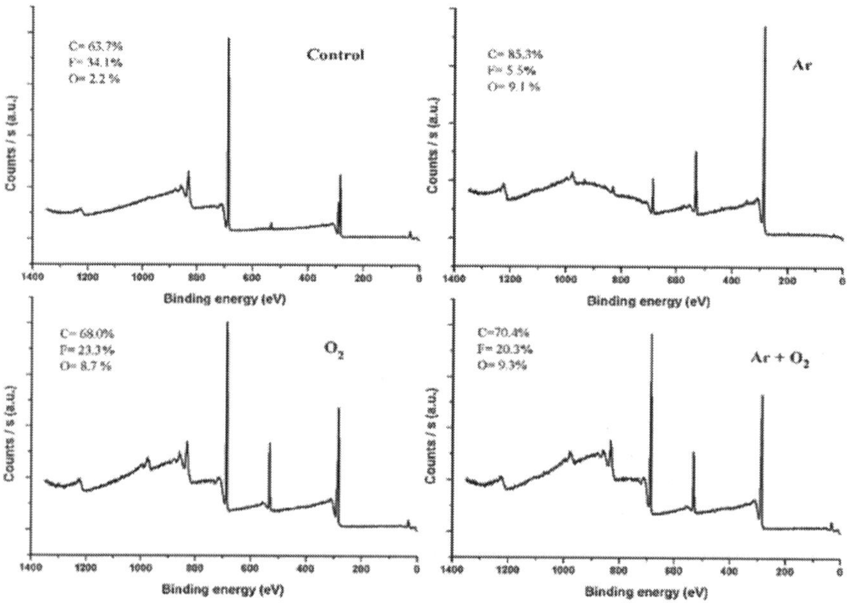

Figure 3. X-ray photoelectron spectroscopy (XPS) survey scan of polyvinylidene fluoride (PVDF) film before and after Ar, O_2 and Ar + O_2 plasma treatment.

3.3. XPS Measurement

XPS measurements were done for control, Ar, O_2 and (Ar + O_2) plasma treated PVDF samples. Carbon, fluorine and oxygen were detected in all the samples as shown in the overall XPS scan in Figure 3 and Table 1. Plasma treatment lowered the fluorine content, however oxygen was increased. There is a large difference in bonding energy between C–F and C–H bonds. The bond energy for F–F bonds is only 1.6 eV, that for H–F bonds 5.85 eV, and that for C–F bonds 4.42 eV [14]. High resolution C1s scans were performed to study the presence of different functional groups on the PVDF surface. C1s peaks were curve fitted to calculate the relative concentration of present functional groups. Three sub peaks C1s peak A, C1s peak B and C1s peak C at 286.4, 291.4 and 288 eV respectively were found. Peak A is assigned to CH_2–CF_2, CHF–CH_2–CHF and O–CH_2 groups. Peak B is assigned to CF_2–CH_2, while peak C is assigned to CH_2–CHF–CH_2, CH_2–CHF–CHF, O–CH_2–CF_2[22]. O 1s peak at 533.7 eV is assigned to C–O [23], while F1s peak at 688.65 eV to C–F bond.

Table 1. Atomic composition of untreated and plasma treated polyvinylidene fluoride (PVDF) film samples.

Sample	C1s Peak A (286.4)	C1s Peak B (291.4)	C1s Peak C (288)	O1s Peak A (533.7)	F1s Peak A (688.65)	C	O	F	O/C	F/C
Control	29.21	16.8	17.7	2.2	34.14	63.7	2.16	34.1	0.03	0.54
Ar	73.42	1.89	10	9.2	5.48	85.3	9.15	5.48	0.11	0.06
O_2	51.54	7.99	8.46	8.7	23.31	68.0	8.7	23.3	0.13	0.34
Ar + O_2	52.14	9.4	8.85	9.3	20.31	70.4	9.31	20.3	0.13	0.29

The untreated PVDF surface showed two main peaks at 291.4 eV (due to CF_2) and 286.4 eV (due to CH_2) components [11,24]. Ar plasma was the most effective in lowering the C1s peak B from 16.8 to 1.89, while O_2 and (Ar + O_2) resulted in moderate levels of de-fluorination (Table 1). The fluorine to carbon ratio (F/C) decreased from 0.54 to 0.06 for Ar and 0.34, 0.29 for O_2 and (Ar + O_2) respectively. Ar plasma also increased the surface oxygen content from 2.16% to 9.15%, however two other plasma gases achieved similar results for oxygen incorporation. The oxygen to carbon ratio (O/C) rose to 0.13 for O_2 and (Ar + O_2) plasma from 0.03 for the control sample. The O/C value was 0.11 for Ar plasma. It is interesting to note the incorporation of oxygen based functional groups into the argon plasma. The oxygen may arise from the following sources; (i) an impurity in Ar gas [25]; (ii) exposure to air after taking out of the plasma chamber before XPS measurement [17].

The C1s spectra of PVDF films before and after RF plasma show that plasma treatment causes de-fluorination and the incorporation of oxygen into the polymer surface (Figure 4). The C1s scans and overall composition (Table 1) indicate that CF_2 carbons were modified into CHF and CH_2 carbons and O–C carbons (O–CH_2 and O–CHF) during plasma exposure as evident by the increase of oxygen content and decrease in C1s Peak A. Increase in hydrophilicity is an important indication of incorporation of oxygen based functional groups. Although Ar plasma resulted in a higher degree of de-fluorination, it did not result in lowering of WCA.

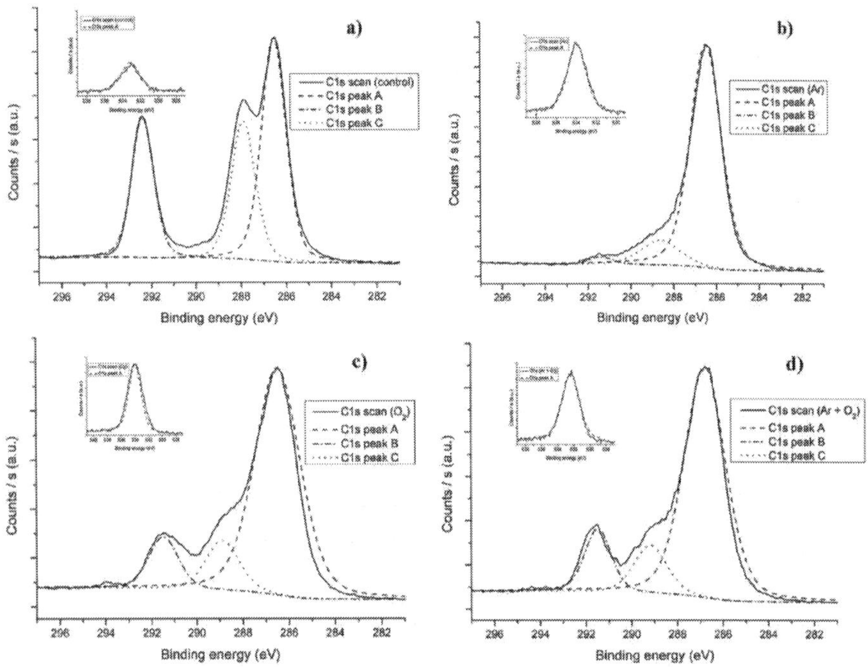

Figure 4. C1s scan spectra of polyvinylidene fluoride (PVDF). (**a**) Control; (**b**) Ar plasma; (**c**) O_2 plasma and (**d**) (Ar + O_2) plasma.

3.4. Abrasion Resistance and Sheet Resistance

PPy coated PVDF film samples abraded by the Martindale abrasion tester were photographed by a microscope (Olympus DP71, Tokyo, Japan) after 0, 200 and 2000 cycles of abrasion (Figure 5). The magnification was set at 10 times for all the images. Some of the coating of the control sample was missing even before the abrasion test due to weak interfacial bonding and nearly all the coating faded from the control sample after 200 and 2000 cycles of abrasion leaving the surface non-conductive.

Plasma treated samples clearly showed better abrasion resistance. Although the Ar plasma sample showed some flaking off after 2000 cycles, it was still significantly better than the control sample. On the other hand the O_2 and (Ar + O_2) plasma treated samples had strong binding as there was only a small visible coating loss. The effectiveness of (Ar + O_2) plasma treatments is clearly evident, as it showed better coating fastness than use of these gases alone. This is due to de-fluorination and incorporation of oxygen based functional groups during plasma treatment [14]. The results showed that Ar plasma resulted in de-fluorination even before abrasion tests. We do not have evidence whether the surface morphology had any variation along with the de-flourination. An AFM investigation of this would be worthwhile in a future study.

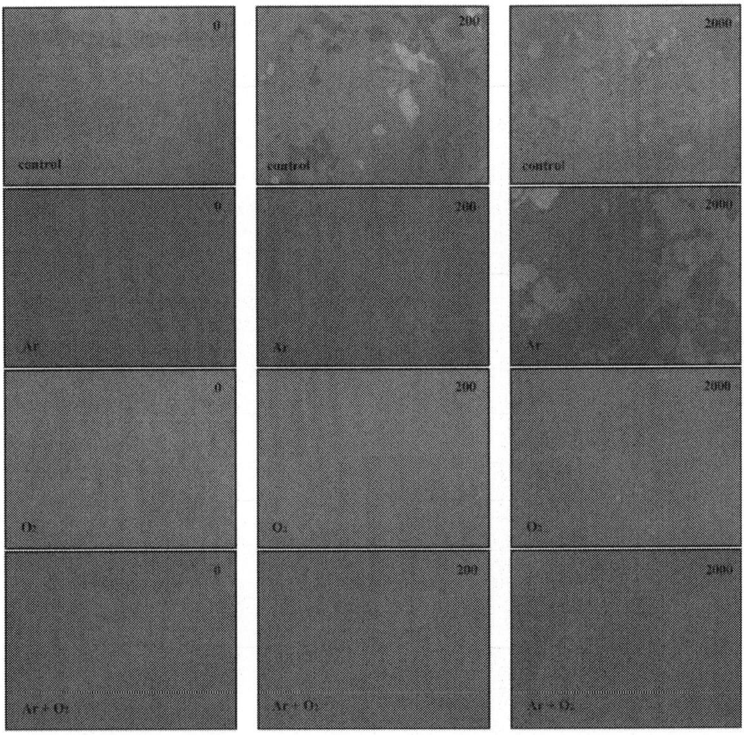

Figure 5. Polypyrrole (PPy) coated polyvinylidene fluoride (PVDF) film sample after 0, 200 and 2000 abrasion cycles in a Martindale abrasion tester for control, Ar, O_2 and (Ar + O_2) samples; magnification 10×.

The Sheet resistance (R_s) results after 0, 200 and 2000 abrasion cycles are presented in Figure 6. Ar and (Ar + O_2) plasma gases lowered the R_s values of the PPy coated PVDF films compared to the control sample (1600 ohms/sq), while O_2 plasma treatment resulted in a small increase in initial R_s values which may be due to O_2 being less affected in de-fluorination compared to Ar and (Ar + O_2). The R_s values could not be calculated for control samples with 200 and 2000 abrasion cycles. The lowest R_s values were noted for (Ar + O_2) gas at 1197 ohms/sq for 120 s plasma treatment, which increased to 2589 ohms/sq after 2000 abrasion cycles. The R_s value of the Ar plasma sample before abrasion was 1452; it rose to 2083 and 66,523 ohms/sq after 200 and 2000 abrasion cycles respectively. Similarly, R_s values for O_2 plasma before abrasion were 2280; they rose to 4218 and 6989 ohms/sq after 200 and 2000 abrasion cycles respectively. The improved sheet resistance as a result of the Ar + O_2 plasma treatment may be attributed to the advantages of both the argon plasma (effective on defluorination) and the oxygen plasma (effective on dehydration and introduction of oxygen functional groups) to achieve improved bonding, hence lower resistivity.

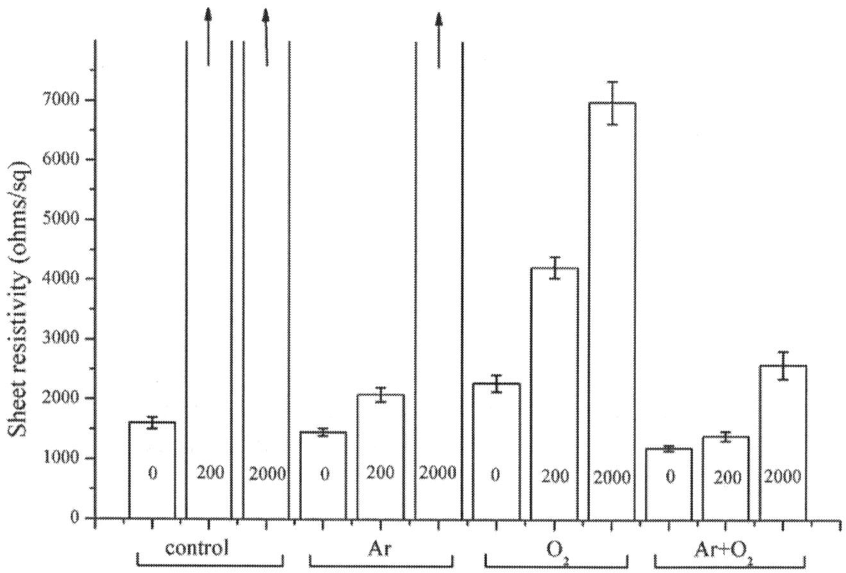

Figure 6. Sheet resistance of polypyrrole (PPy) coated polyvinylidene fluoride (PVDF) samples; control, Ar, O_2 and ($Ar + O_2$).

The sheet resistance data of PPy coated PVDF samples in Figure 6 and photographs of abraded samples in Figure 5 agree with each other and all confirm the order of plasma gas effectiveness as ($Ar + O_2$) > O_2 > Ar. Contact angle measurements show an improvement in the PVDF surface wettability after the plasma treatment. The decrease of the F/C ratio and water contact angle prove that both chemical and physical effects play an important role in the adhesion.

It is interesting to note that control, Ar and O_2 plasma treated PVDF samples had a grainy surface morphology which did not lower the R_s value as much as the ($Ar + O_2$) plasma treated surface having a flat yet nano-sized surface roughness, a similar trend was observed for the bonding strength. Both XPS and surface imaging indicates that incorporation of oxygen functional groups and nano sized surface roughness are important for coating fastness. This was confirmed by the connection of abrasion resistance imaging, sheet resistance measurement and the chemical composition data in Table 1. Similar results were obtained in our previous work on polyester fabric [19].

The treatment of PVDF using oxygen, argon and hydrogen plasmas were reported in a similar study by Park *et al.* [26], where they reported using remote plasma treatment in which the sample was situated in the after-glow area and the authors focussed on the individual effects of the above mentioned plasma gases. Whereas, in our experiments the sample was placed inside the plasma source and the effects of argon (inert gas), oxygen (reactive gas) and the combination argon + oxygen plasma were used to enable a better understanding of the role of argon and oxygen plasma and the effect of the combination. Overall, results showed that the most effective treatment was the argon + oxygen combination, followed by oxygen and argon. Our results agree with that of Park *et al.* [26] in the contact angle data for oxygen and argon plasma, that is, argon plasma treatment was more effective than oxygen in hydrophilicity manifesting as a decrease in contact angle. Similarly, oxygen plasma was not very effective in defluorination, supporting the earlier study by Park *et al.* [26]. In this study we were able to use the advantages of both argon plasma (effective on defluorination) and oxygen plasma (effective on dehydration and introduction of oxygen functional group) to achieve improved bonding. In other words, the improved bonding of the combination may be attributed to the combined effect. Also the argon plasma produced more reactive sites on the substrate surface, which would enhance the binding of functional groups.

4. CONCLUSIONS

We have studied the improvement of binding of polypyrrole with a PVDF thin film using low pressure plasma. Ar, $O_2$and Ar + O_2 gases were used in the plasma treatment. The abrasion data suggest adequate strength of bonding with the PVDF surface after plasma treatment. The plasma induced increase in hydrophilicity and surface functionalization with oxygen based groups contributed to the improved coating adhesion and fastness. These results clearly show the effectiveness of Ar + O_2 plasma in lowering the resistivity and improving the binding strength of the PPy coating.

ACKNOWLEDGMENTS

We wish to acknowledge Gayathri Devi Rajmohan and Zhiqiang Chen for help with the plasma experiments, Johan du Plessis and RMIT for access to the XPS facility and the Higher Education Commission of Pakistan for a PhD scholarship.

REFERENCES

1. Drean, E.; Schacher, L.; Adolphe, D.; Bauer, F. Smart Textiles for Automotive: Application to Airbag Development. In*4th International Workshop on Wearable and Implantable Body Sensor Networks (BSN 2007)*; Leonhardt, S., Falck, T., Mähönen, P., Eds.; Springer Berlin-Heidelberg: Aachen, Germany, 2007; Volume 13, pp. 155–160.
2. Moussaif, N.; Pagnoulle, C.; Riga, J.; Jérôme, R. XPS analysis of the PC/PVDF interface modified by PMMA. Location of the PMMA at the interface. *Polymer* **2000**, *41*, 3391–3394.
3. Sencadas, V.; Gregorio, R.; Lanceros-Méndez, S. α to β phase transformation and microestructural changes of PVDF films induced by uniaxial stretch. *J. Macromol. Sci. Part B Phys.* **2009**, *48*, 514–525.
4. Chang, J.; Dommer, M.; Chang, C.; Lin, L. Piezoelectric nanofibers for energy scavenging applications. *Nano Energy* **2012**, *1*, 356–371.
5. Kim, H.; Kim, J.-H.; Kim, J. A review of piezoelectric energy harvesting based on vibration. *Int. J. Precis. Eng. Manuf.***2011**, *12*, 1129–1141.
6. Hutchison, A.S.; Lewis, T.W.; Moulton, S.E.; Spinks, G.M.; Wallace, G.G. Development of polypyrrole-based electromechanical actuators. *Synth. Met.* **2000**, *113*, 121–127.
7. Choi, S.; Jiang, Z. A novel wearable sensor device with conductive fabric and PVDF film for monitoring cardiorespiratory signals. *Sens. Actuators A Phys.* **2006**, *128*, 317–326.
8. Pascu, M.; Debarnot, D.; Durand, S.; Poncin-Epaillard, F. Surface Modification of PVDF by Microwave Plasma Treatment for Electroless Metallization. In *Plasma Processes and Polymer*; Wiley-VCH Verlag GmbH & Co. KGaA: Taormina, Italy, 2005; pp. 157–176.
9. Wallace, G.G.; Campbell, T.E.; Innis, P.C. Putting function into fashion: Organic conducting polymer fibres and textiles. *Fibers Polym.* **2007**, *8*, 135–142.
10. Bhat, N.V.; Geetha, P.; Pawde, S.; Nallathambi, R. Preparation of poly(vinylidene fluoride)–polypyrrole composite films by electrochemical synthesis and their properties. *J. Appl. Polym. Sci.* **1995**, *58*, 2251–2257.
11. Mansouri, J.; Burford, R.P. Characterization of PVDF-PPy composite membranes. *Polymer* **1997**, *38*, 6055–6069.
12. Sun, J.; Yao, L.; Gao, Z.; Peng, S.; Wang, C.; Qiu, Y. Surface modification of PET films by atmospheric pressure plasma-induced acrylic acid inverse emulsion graft polymerization. *Surf. Coat. Technol.* **2010**, *204*, 4101–4106.
13. Keller, M.; Ritter, A.; Reimann, P.; Thommen, V.; Fischer, A.; Hegemann, D. Comparative study of plasma-induced and wet-chemical cleaning of synthetic fibers. *Surf. Coat. Technol.* **2005**, *200*, 1045–1050.

14. Park, Y.W.; Inagaki, N. A new approach for selective surface modification of fluoropolymers by remote plasmas. *J. Appl. Polym. Sci.* **2004**, *93*, 1012–1020.
15. Yamada, Y.; Yamada, T.; Tasaka, S.; Inagaki, N. Surface modification of poly(tetrafluoroethylene) by remote hydrogen plasma. *Macromolecules* **1996**, *29*, 4331–4339.
16. Crowe, R.; Badyal, J.P.S. Surface modification of poly(vinylidene difluoride)(PVDF) by LiOH. *J. Chem. Soc. Chem. Commun.* **1991**, *1991*, 958–959.
17. Duca, M.D.; Plosceanu, C.L.; Pop, T. Surface modifications of polyvinylidene fluoride (PVDF) under rf Ar plasma.*Polym. Degrad. Stab.* **1998**, *61*, 65–72.
18. Momose, Y.; Noguchi, M.; Okazaki, S. Ar, O_2 and CF_4 plasma treatment of poly-(vinylidene fluoride), polyimide and polyamidoimide and its relationship to wettability. *Nucl. Instrum. Methods Phys. Res. Sect. B Beam Interact. Mater. At.***1989**, *39*, 805–808.
19. Mehmood, T.; Dai, X.J.; Kaynak, A.; Kouzani, A. Improved bonding and conductivity of polypyrrole on polyester by gaseous plasma treatment. *Plasma Process. Polym.* **2012**, *9*, 1006–1014.
20. Dai, X.J.; Chen, Y.; Chen, Z.; Lamb, P.R.; Li, L.H.; du Plessis, J.; McCulloch, D.G.; Wang, X. Controlled surface modification of boron nitride nanotubes. *Nanotechnology* **2011**, *22*.
21. Bautista, K. Four-Point Probe Operation. Available online: http://www.utdallas.edu/research/cleanroom/manuals/documents/4pointFinal.pdf (accessed on 30 April 2004).
22. Vandencasteele, N.; Reniers, F. Plasma-modified polymer surfaces: Characterization using XPS. *Relat. Phenom.* **2010**,*178–179*, 394–408.
23. Pawde, S.M.; Deshmukh, K. Surface characterization of air plasma treated poly vinylidene fluoride and poly methyl methacrylate films. *Polym. Eng. Sci.* **2009**, *49*, 808–818.
24. Bodenes, L.; Naturel, R.; Martinez, H.; Dedryvère, R.; Menetrier, M.; Croguennec, L.; Pérès, J.-P.; Tessier, C.; Fischer, F. Lithium secondary batteries working at very high temperature: Capacity fade and understanding of aging mechanisms. *J. Power Sources* **2013**, *236*, 265–275.
25. Dai, X.; Hamberger, S.; Bean, R. Reactive plasma species in the modification of wool fibre. *Aust. J. Phys.* **1995**, *48*, 939–952.
26. Park, Y.W.; Inagaki, N. Surface modification of poly(vinylidene fluoride) film by remote Ar, H_2, and O_2 plasmas.*Polymer* **2003**, *44*, 1569–1575.

Chapter 7

Abrasion, Erosion and Cavitation Erosion Wear Properties of Thermally Sprayed Alumina Based Coatings

*Ville Matikainen *, Kari Niemi, Heli Koivuluoto and Petri Vuoristo*

Department of Materials Science, Tampere University of Technology, Korkeakoulunkatu 6, FI-33720 Tampere, Finland

ABSTRACT

Thermally-sprayed alumina based materials, e.g., alumina-titania (Al_2O_3-TiO_2), are commonly applied as wear resistant coatings in industrial applications. Properties of the coatings depend on the spray process, powder morphology, and chemical composition of the powder. In this study, wear resistant coatings from Al_2O_3 and Al_2O_3-13TiO_2 powders were sprayed with plasma and high-velocity oxygen-fuel (HVOF) spray processes. Both, fused and crushed, and agglomerated and sintered Al_2O_3-13TiO_2 powders were studied and compared to pure Al_2O_3. The coatings were tested for abrasion, erosion, and cavitation resistances in order to study the effect of the coating structure on the wear behavior. Improved coating properties were achieved when agglomerated and sintered nanostructured Al_2O_3-13TiO_2 powder was used in plasma spraying. Coatings with the highest wear resistance in all tests were produced by HVOF spraying from fused and crushed powders.

Keywords: abrasion; erosion; cavitation; plasma spray; high-velocity oxygen-fuel (HVOF); alumina; alumina-titania

1. INTRODUCTION

Metal components are the basis of engineering applications and constructions and are exposed to wide range of conditions and environments, e.g., in process industries. The ductile metallic base material can be subjected to different types of wear by the surrounding operation environment. In such cases, hard thermal spray coatings can be utilized to improve the wear resistance of the metallic surface [1]. Alumina (Al_2O_3) and alumina-titania (Al_2O_3-TiO_2) materials are widely used as thermally-sprayed protective coatings against abrasion, erosion, and cavitation erosion wear, e.g., in textile, pulp and paper, and pump industries [2]. Alumina is a hard and relatively cheap material with high hardness, good wear, and corrosion and thermal resistance [2,3]. It has also high electric resistance, which has made Al_2O_3 attractive for coating applications where electric insulation and high breakthrough voltage is sought. Fracture toughness of Al_2O_3 coatings can be increased by small addition of TiO_2, which is mechanically blended with Al_2O_3 particles or mechanically cladded on the Al_2O_3 particles as a layer of small TiO_2 particles [1]. As a down side, the added TiO_2 also decreases hardness and wear resistance of the coating [2]. During the last decade, nanostructured powders have been gaining increasing attention and improved coating properties, e.g., higher toughness, adhesion, and wear resistance, have been reported [4,5,6,7,8]. Typically, Al_2O_3 and Al_2O_3-TiO_2 powders have been manufactured by blending Al_2O_3 and TiO_2 fused and crushed powders or by cladding Al_2O_3 particles with TiO_2. Manufacturing of the nanostructured spherical particles with good flowability is most commonly done by spray drying. The powder manufacturing method and TiO_2 distribution in the powder and sprayed coating has been studied and found to affect the coating properties [8,9,10]. Evenly distributed TiO_2 in the spray-dried powders improves the coating properties compared to coatings sprayed using blended or mechanically clad powders resulting in distinguishable areas of Al_2O_3 and TiO_2 [8,10].

The Al_2O_3 coatings are typically sprayed with plasma torch, which is capable of melting generally any material with sufficient difference between the material's melting and vaporization (or decomposition) temperature [11]. The particle size distribution of oxide ceramic powders used in conventional plasma spraying is often +45-22 μm. Plasma formation related arc fluctuation and different particle sizes combined with radial powder feeding result in wide range of particle trajectories and thermal histories. This scattering causes defects in the coating structure, e.g., partially molten particles, unmolten particles, and related porosity. One approach to improve the Al_2O_3 coatings has been the use of high-velocity oxygen fuel (HVOF) spray process, which is based on combustion of oxygen and fuel gas or liquid. The melting temperature of Al_2O_3 is 2040 °C, which makes it possible to be sprayed also by combustion based spray processes, e.g., flame spraying and HVOF spraying [12,13,14,15,16]. Maximum flame temperatures between 2828 and 3160 °C can be achieved depending on the used fuel gas [17]. Very dense coatings can be produced by HVOF spray processes compared to plasma-sprayed coatings due to the higher particle velocity. In general, smaller particle size is necessary for sufficient melting of the particles in the HVOF process compared to plasma spraying. The melting and deposition efficiency of HVOF-sprayed ceramics have been regarded as low, but this has been overcome by the use of ethylene as the fuel gas [17]. The requirement for spraying Al_2O_3, and other ceramics, with HVOF processes, is the sufficient particle flight time and heating from the powder injection to the point where the gas jet temperature drops below the particle temperature. This has been achieved by using HVOF torches with the powder injection located in the combustion chamber [2].

Typically, the Al_2O_3 coatings are produced with plasma spraying. During recent years, the coating properties have been improved by the use of HVOF processes, as well as with the development of nanostructured powders for plasma spraying,*i.e.*, agglomerated and sintered powders from nanosized primary particles. Both, nanostructured coatings and HVOF-sprayed coatings have been reported to possess significantly higher wear resistance compared to conventional plasma-sprayed coatings [4,5,6,7,8,18,19]. However, the performance of nanostructured plasma-sprayed coatings and HVOF-sprayed ceramic coatings has not been truly compared under abrasive, erosive, and cavitation erosion wear. The cavitation erosion of thermally-sprayed coatings has been studied increasingly during the last decades, but most of the studies have focused on metallic and carbide coatings [20,21,22,23], while ceramic materials, and especially ceramic coatings, have not been studied as actively. Some cavitation erosion studies of bulk ceramics have been published [24,25,26], but only few studies concerning thermally-sprayed ceramic coatings [27,28,29]. Typically, these materials have a brittle structure and lower cohesion, which results in considerably lower resistance against cavitation erosion as compared to, e.g., carbide coatings. The coating properties provided by HVOF spraying and nanostructured powder materials, however, are potential candidates to improve the performance of ceramic coatings in cavitation erosion environments.

Alumina-based coatings are applied to various applications, in which they are exposed to different types of wear, e.g., abrasion, erosion, or cavitation erosion wear. Especially, the current research on erosion wear properties of HVOF-sprayed ceramic coatings and cavitation erosion wear of ceramic coatings is very limited. For this reason, it is important to study the wear behavior of ceramic coatings under solid particle erosion and cavitation erosion. By comparing the results to abrasion wear behavior, a good understanding of the overall coating performance can be drawn. In this study, Al_2O_3 and Al_2O_3-

13TiO$_2$ powders with different morphologies (agglomerated and sintered *vs.* fused and crushed) were sprayed with atmospheric plasma spray (APS) and high-velocity oxygen fuel (HVOF) spray systems. Structures and the wear behavior of these coatings were studied under abrasion, erosion, and cavitation erosion wear conditions.

2. EXPERIMENTAL TECHNIQUES

2.1. Coating Manufacturing

Alumina-based coatings were produced by atmospheric plasma spraying (APS) and high-velocity oxygen-fuel (HVOF) spraying. APS coatings were sprayed with Plasma Technik A-3000S 4/2 plasma spray system equipped with a conventional F4MB (Sulzer Metco AG, Switzerland) plasma torch using a radial powder injection. Argon and hydrogen were used as process gases. HVOF coatings were sprayed with TopGun (GTV Gmbh, Germany) spray gun. A variety of process gasses can be used with TopGun and it has 22 mm long combustion chamber, which makes it especially capable of spraying ceramic powders. Ethylene was used as fuel gas for the HVOF spraying. Spray parameters for APS and HVOF spray processes are presented in Table 1 and Table 2, respectively. Coatings were sprayed on grit-blasted (Al_2O_3 grits, 36 Mesh) 20 mm × 100 mm × 5 mm low carbon steel (Fe52) substrates.

Table 1. Spray parameters for plasma spraying (APS).

Spray Parameter	Powder	
	Al_2O_3	Al_2O_3-13TiO_2
Current [A]	600	530
Argon [slpm]	41	41
Hydrogen [slpm]	14	14
Nozzle diameter [mm]	6	6
Spray distance [mm]	120	120
Powder feed rate [g/min]	30	30

2.2. Powder Materials

Table 3 shows the information of the powders used in this study. Fused and crushed Al_2O_3 powders were sprayed with APS and HVOF spray processes. Particle sizes were optimized for both processes. Coarser powders were used for plasma spraying and the finer powders for HVOF spraying. In addition, two different powder types of Al_2O_3-13TiO_2 were used in both spray processes (with optimized particle sizes). Al_2O_3-13TiO_2 powders were either fused and crushed or agglomerated and sintered. Agglomerated and sintered powders, having the nanostructures, consist of small evenly distributed primary particles of Al_2O_3 and TiO_2. These powders have spherical powder morphology, which is presented in Figure 1a. The primary particle size after manufacturing process is approximately from 50 to 500 nm, with few larger particles up to 1500 nm. The conventional powders of Al_2O_3 and Al_2O_3-TiO_2 are manufactured by fusing and crushing method, resulting in dense irregular blocky particles. The powder morphology of fused and crushed Al_2O_3-TiO_2 blended powder is presented inFigure 1b. Some lighter grey TiO_2 particles can be observed in Figure 1b, while darker grey particles are Al_2O_3.

Table 2. Spray parameters for high-velocity oxygen fuel (HVOF).

Spray parameter	Powder Al_2O_3, Al_2O_3-13TiO_2
Ethylene [slpm]	93
Oxygen [slpm]	270
Spray distance [mm]	150
Chamber/nozzle [mm]	22/135
Powder feed rate [g/min]	40

Table 3. Information of the used powders and sample names, F/C stands for fused and crushed, and A/S for agglomerated and sintered.

Sample name	Chemical Composition	Manufacturing method	Manufacturer	Trade name	Particle size [µm]	Spray method
APS_A	Al_2O_3	F/C	H.C. Starck, Goslar, Germany	Amperit 740.1	-45+22	Plasma
HVOF_A	Al_2O_3	F/C	H.C. Starck, Goslar, Germany	Amperit 740.8	-20+5	HVOF
APS_AT1	Al_2O_3-13TiO_2	F/C	Sulzer Metco, Wohlen, Switzerland	Amdry 6228	-45+22	Plasma
HVOF_AT1	Al_2O_3-13TiO_2	F/C	Sulzer Metco, Wohlen, Switzerland	Amdry 6220	-22+5	HVOF
APS_AT2	Al_2O_3-13TiO_2	A/S	Millidyne, Tampere, Finland	Neoxid A103	-45+9	Plasma
HVOF_AT2	Al_2O_3-13TiO_2	A/S	Millidyne, Tampere, Finland	Neoxid A103	-32+5	HVOF

Figure 1. Powder morphologies of (**a**) agglomerated and sintered Al_2O_3-13TiO_2powder (Neoxid A103, HVOF cut) and (**b**) fused and crushed Al_2O_3-13TiO_2 powder (Amdry 6228, APS cut). SEM images.

2.3. Characterisation Techniques

Coating microstructures were analyzed using a field-emission scanning electron microscope (FE-SEM), (ULTRAplus, Carl Zeiss Microscopy GmbH, Germany). Structures were studied from metallographic cross-sectional coating samples. General coating structures, powder morphologies, and worn surfaces after wear tests were characterized with a scanning electron microscope, SEM (XL30, Philips, The Netherlands). Furthermore, mechanical behavior of the coatings was investigated by microhardness measurements. Vickers hardnesses (HV0.3) were measured as an average of 10 measurements using a microhardness tester (MMT-X7, Matsuzawa, Japan).

2.4. Wear Tests

Abrasion wear behavior of the coatings was evaluated using a modified version of the ASTM G 65 dry rubber-wheel abrasion test, where five samples could be tested simultaneously. The total duration of the test was 60 min. Weight losses were measured after each 12-minute period with an accuracy of 0.001 g. The surface speed of the rubber wheel was 1.64 m/s, which resulted in a total wear length of 5900 m. Each sample was pressed against the rubber wheel with a force of 13 N. The test simulates three-body abrasion condition, where dry quartz sand (SiO_2) with a grain size of 0.1–0.6 mm was used as an abrasive (Figure 2). The flow rate of the abrasive was 25 g/min. The results are given as an average of three measurements. The sample surfaces were fine-grinded (1200 SiC paper) to remove the surface roughness of the as-sprayed coatings.

Figure 2. SEM image of the quartz sand used in abrasion and erosion wear tests.

Erosion wear behavior of the coatings was studied with a centrifugal erosion tester. Total amount of 12 samples were attached simultaneously in the sample holders, located on the outer ring in fixed positions. Sample holders, with angles of 30 and 90 degrees, were used. Rotation speed was set to 6000 rpm for the test and 1 kg of the same quartz sand as in abrasion wear test was used as the erosive. The sample surfaces were fine-grinded (1200 SiC paper) before tests.

Cavitation erosion tests were performed with an ultrasonic transducer (VCX-750, Sonics & Materials, USA) according to the ASTM G32-10 standard [30]. The optional stationary specimen method described in the standard was used. Schematic presentation of the test setup is shown in Figure 3a. The coating surfaces to be tested were grinded flat and polished with a polishing cloth and diamond suspension (3 μm). Samples were cleaned in ultrasonic bath with ethanol and weighed after drying. Samples were attached on a stationary sample holder and the head of the ultrasonic transducer was placed at the distance of 0.5 mm. Samples were weighed after 5, 15, 30, 60, 90, and 120 min. Depending on the wear rate, samples were tested further for a total of 150 or 210 min. A replaceable tip made of Ti-6Al-4V was used as the cavitation resistant horn piece (Figure 3b).

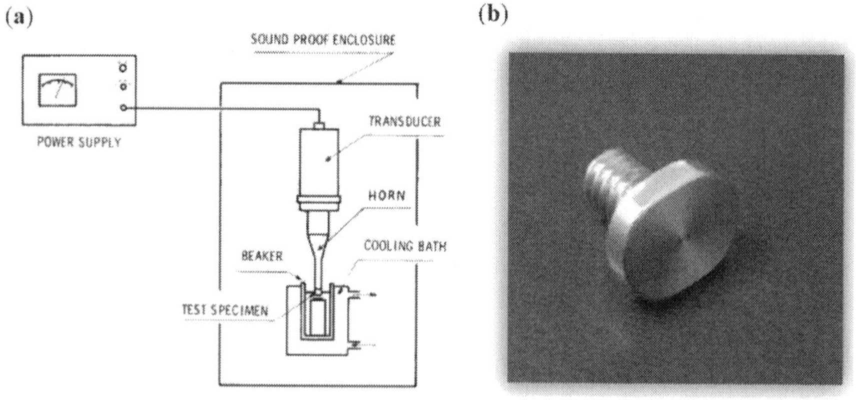

Figure 3. Cavitation erosion test (**a**) schematic presentation of test setup, modified from ASTM G32-10 standard [29] and (**b**) the replaceable Ti-6Al-4V tip used.

3. RESULTS AND DISCUSSION

3.1. Coating Microstructure

The cross-sections of the coatings were studied with FE-SEM. General views of the coating structures are presented in Figure 4 (APS coatings in 4a–c and HVOF coatings in 4d–f). All coatings are relatively dense. In addition, all HVOF-sprayed coatings were denser compared to their plasma counterparts. The coatings sprayed from fused and crushed powders showed typical coating structure with evenly distributed porosity. The HVOF_AT2 coating showed more large pores compared to other HVOF coatings. The porosity was found to result from pull-outs and removal of completely unmelted particles in the structure.

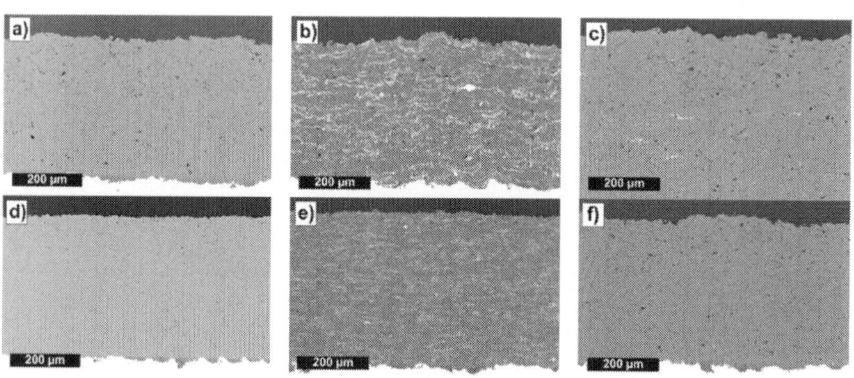

Figure 4. Structure of (**a**) APS_A; (**b**) APS_AT1 (F/C); (**c**) APS_AT2 (A/S); (**d**) HVOF_A; (**e**) HVOF_AT1 (F/C); and (**f**) HVOF_AT2 (A/S) coatings. SEM images.

The detailed microstructures of APS_A and APS_AT1 (F/C) coatings are presented in Figure 5. Both coatings show a quite distinctive vertical microcracking of the lamellas throughout the coating resulting from locally insufficient bonding and relaxation of quenching stresses [31]. The thickness of the white TiO_2 lamellas in the coating sprayed from the blended powder ranges from submicron thickness to a few micrometers (Figure 5b). The adherence of the TiO_2 lamellas in the structure seems to be good and there is no observable cracking particularly related to TiO_2 splats.

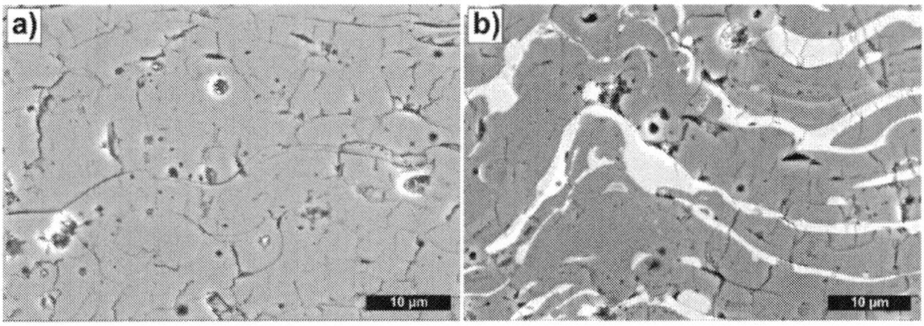

Figure 5. Microstructure of (**a**) APS_A and (**b**) APS_AT1 coatings. FE-SEM images.

Very dense HVOF coatings were produced from the fused and crushed powders. Both HVOF_A (Figure 6a) and HVOF_AT1 (Figure 6b) coatings have a few visible, insufficiently melted, rounded Al_2O_3 particles in the structure. The TiO_2particles in Figure 6b seem to have melted completely during the spray process. Both coatings show some vertical cracking, but limited cracking of splat interfaces as compared to plasma-sprayed coatings. Structural details of all HVOF-sprayed coatings were smaller compared to APS coatings due to the smaller particle size distribution used. In addition, the higher velocity involved in the HVOF process promoted the more complete filling of the surface roughness during the spraying. This can be seen from the filling of the surface around the partially melted Al_2O_3 particles in both coatings. It can be speculated if the smaller size of the particles also leads to lower quenching stresses as a result of smaller flattened particle size and larger number of splats sharing the stresses forming in the coating. The smaller diameter of the splat would mean smaller quenching stresses if full contact and good adhesion between flattened particle and surface is expected. In addition, the high velocity impact of the small Al_2O_3 particles have been reported reaching extremely high cooling rates, which can promote the formation of nanocrystalline γ-Al_2O_3 phase or even an amorphous structure [32]. Furthermore, Sakoda *et al.*[33] have reported that the hardness values of different structures in Al_2O_3 coatings vary between α-Al_2O_3, γ-Al_2O_3 and nanocrystalline phases.

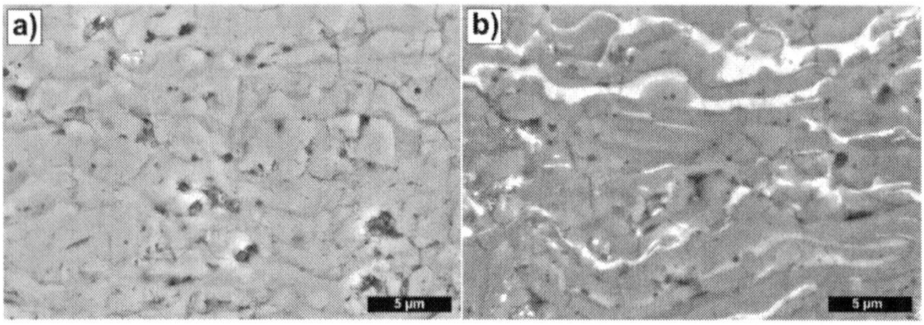

Figure 6. Microstructure of (**a**) HVOF_A and (**b**) HVOF_AT1 coatings. FE-SEM images.

When comparing the structures of APS- and HVOF-sprayed coatings from agglomerated and sintered Al_2O_3-13TiO_2powders (APS_AT2 and HVOF_AT2), they showed unique structure with partially molten composite particles. A partially molten particle from APS_AT2 coating is visible in Figure 7a and it consists of lighter and darker grey areas. The color difference arises from the variation of TiO_2 content within the partially molten particle. TiO_2 melts at lower temperature (1854 °C) than pure Al_2O_3 (2040 °C) [7] and has been suggested of existing as solid solution with γ-Al_2O_3 [9]. The lighter segregated areas were found to contain more TiO_2 than the darker areas, based on EDS analysis. This sort of partially molten structure has higher viscosity than fully molten particles, which presents a risk of increased coating porosity during plasma spraying. These particles are not able to flatten as well as fully molten particles, which can lead to insufficient filling of surface roughness. Similar composite structures exist in the HVOF-sprayed coating in Figure 7b. It is noteworthy that the flattening and bonding of the partially molten areas seems better compared to APS-sprayed coating, *i.e.*, lamella interfaces are less visible. Despite the dense structure, regions of completely unmolten particles were also found in the detailed structural characterization of the HVOF_AT2 coating. These particles showed clearly the loose structure of the powder primary particles. The reason for the existence of unmolten particles in the coating is most likely caused by the large particles that do not even have the sufficient time to be partially melted from the inside. The surface of such particle starts melting during the spray process, but the melt front does not have the time to reach the insides of the particle. In addition, porosity inside the agglomerated and sintered particle can slow down the melting [34].

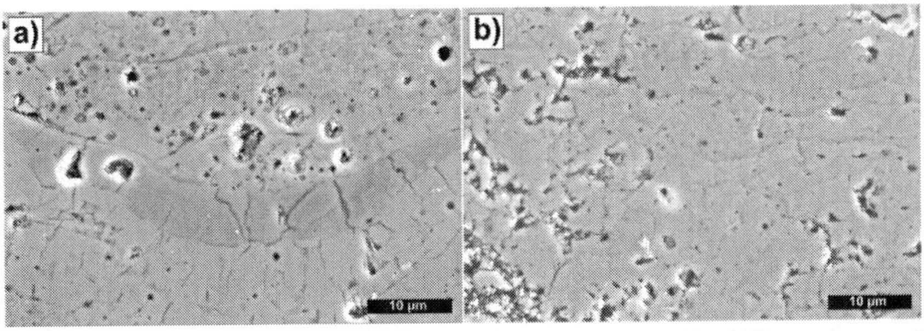

Figure 7. Structural details: (**a**) Partially molten area in APS_AT2 coating; and (**b**) partially molten and unmolten areas in the HVOF_AT2 coating. FE-SEM images.

3.2. Hardness Measurements

The Vickers hardness values (HV0.3) of the coatings are displayed in Table 4. Both coatings sprayed from the agglomerated and sintered powders show slightly lower hardness values compared to the coatings sprayed from fused and crushed powders. There was not any significant difference in the hardness values between the Al_2O_3 and Al_2O_3-TiO_2coatings sprayed from fused and crushed powders.

Table 4. Vickers hardness values (HV0.3) and standard deviations of the coatings.

Measured quantity	APS_A	HVOF_A	APS_AT1	HVOF_AT1	APS_AT2	HVOF_AT2
Hardness [HV0.3]	1105	1117	1089	1075	1009	1027
Standard deviation	171	54	102	58	142	81

3.3. Cavitation Erosion Wear Resistance

Laboratory scale accelerated test carried out with an ultrasonic transducer is a fast way of testing the material's resistance to cavitation erosion [30]. The tests ability to mimic most real application conditions can be argued, but it provides standard conditions where the materials erosion resistance to bubble collapse induced shocks, i.e., cavitation erosion, can be tested. Indeed, the test can be thought as if the coating surface was continuously bombarded by shock waves produced by the numerous cavitation bubbles collapsing near the sample surface. The bubble size without the sample surface can be expected to be approximately 10 µm [35].

The graph in Figure 8 shows the weight losses of the samples as a function of time. It can be noticed that all the samples have linear weight loss, i.e., constant wear rate, throughout the test. For bulk metallic samples, the cavitation erosion wear rate curve typically includes an incubation period in the beginning, during which the metal surface is plastically deformed and no weight loss is observed. After this acceleration stage takes place and the wear rate of the metallic sample starts increasing [36]. This, however, is not the case with brittle ceramic coatings. Incubation stage cannot be observed as wear begins immediately by the removal of the poorly adhered particles from the surface. In addition, the acceleration phase of ceramic coatings seems to take place during the very first minute. This difference between sprayed coatings and bulk materials was also noted by Schwetzke and Kreye [28] in their study of HVOF coatings. The linear wear rates of the samples observed in the graph indicate that the maximum wear rate is reached during the very first minutes.

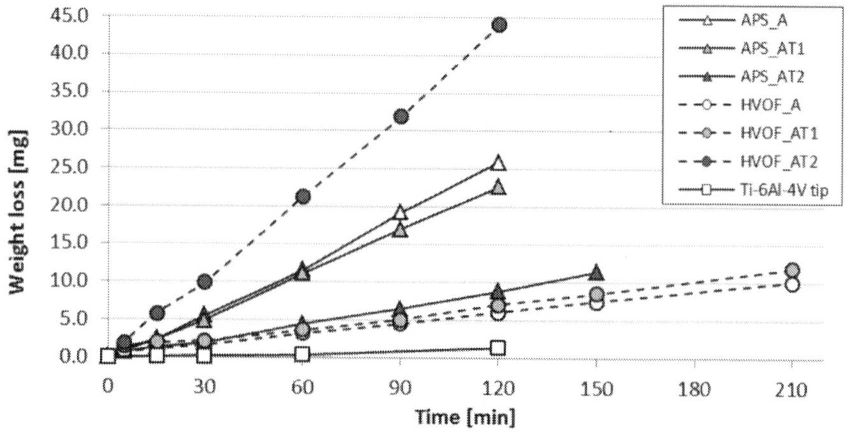

Figure 8. Weight losses of APS and HVOF coatings and Ti-6Al-4V tip material *versus*time in the cavitation erosion test.

The highest cavitation erosion resistance was observed on the coatings HVOF_A, HVOF_AT1 and APS_AT2 with 3.0, 3.4 and 4.6 mg/h wear rate, respectively. Both, APS_A and APS_AT1, plasma-sprayed coatings eroded faster with wear rates of 13.0 and 11.4 mg/h. The highest wear rate was measured for HVOF_AT2 coating (22.1 mg/h). The very good performance of APS_AT2 coating was due to improved coating cohesion by the evenly distributed TiO_2 in the powder and coating. Overall, the cavitation erosion test gives very good information about the coating structure and cohesion as the coating defects are the major factors accelerating the erosion.

The heterogeneous structure of the thermally-sprayed oxide coating consists of (1) well melted structure, (2) partially or unmelted particles, (3) lamella interfaces, (4) porosity, and (5) microcracks [37]. The reason for observed linear wear behavior of all the coating samples derives from the coating structure and material properties. As cavitation erosion begins, the shock waves created by the collapse of the cavitation bubbles near the surface start working the whole surface area. Poorly bonded particles or lamellas on the polished surfaces are the first ones to be removed by the cavitation erosion. Simultaneously, the well-melted areas of the coating surface start undergoing cavitation erosion. Typically, good cavitation erosion resistance is linked to material ductility and cohesion [36], which could favor the composite structures and evenly distributed TiO_2 observed in the coatings produced from agglomerated and sintered powders.

Single lamella in the coating consists of microcracks as shown in Figure 9a. These microcracks are caused by the quenching stresses during the cooling of the lamella. The weak areas mostly worked by the cavitation erosion are the visible edges of the lamella as well as the microcrack network. The wear mechanism of single lamella can be seen in Figure 9b and the material removal seems to take place purely by brittle fracture. The coating's characteristic columnar structure is being picked apart, especially from the edges and the microcracks. A closer image of material loss from the microcracks is presented in Figure 9c. The arrow points to a small particle about to be fractured and removed at the intersection of microcracks. When the microcracks of the lamella have propagated and connected with other cracks, larger pieces are removed. This is further promoted by the fact that the lamellas of plasma-sprayed coating are only partially in contact with the underlying surface. As thc crosion wear progresses, the surface becomes rougher and takes a more blocky form. The small-scale erosion of the lamella is not, however, the main mechanism of material removal. Major fraction of the coating removal takes place in larger particles and the controlling factor of the erosion rate is the overall strength of the coating and the amount of defects, e.g., splat interfaces. Most of the material is lost in larger blocky particles, as a result of crack propagation through quenching cracks and poor interfaces. This mechanism typically involves several splats simultaneously. The debris particle size is smaller for coatings with good cohesion and small-scale structure, e.g., HVOF-sprayed coatings from fused and crushed powders, whereas very large debris particles can be observed for coatings with large amounts of poorly melted particles and low cohesion. Some debris from the agglomerated and sintered coating of Al_2O_3-13TiO_2 (APS_AT2), containing partially molten particles, is presented in Figure 10a. The rough surface of the particle shows the partially molten structure. The structure consists of Al_2O_3 regions in TiO_2-rich matrix. Similar structure from the cross section sample was presented in Figure 7a, showing the same granular texture. A debris particle from a well-melted region of the same coating is presented in Figure 10b. The debris particle shows blocky surface of a brittle fracture and the columnar structure of the lamellas is visible. The top and bottom sides of the particle show larger areas of smooth splat surface. When the cohesion of the coating is poor and the structure includes large amounts of weak surfaces, e.g., splat interfaces, the debris particle size and shape change drastically from the well-melted particle seen in Figure 10b. The HVOF_AT2 coating showed poor performance in cavitation erosion tests and debris particle from the coating can be seen inFigure 10c. The size of the particle is large, approximately 60–100 μm, and the shape is rough but not blocky as would be the case if cracking through splats took place. The failure has progressed through the weak boundaries of insufficiently melted particles.

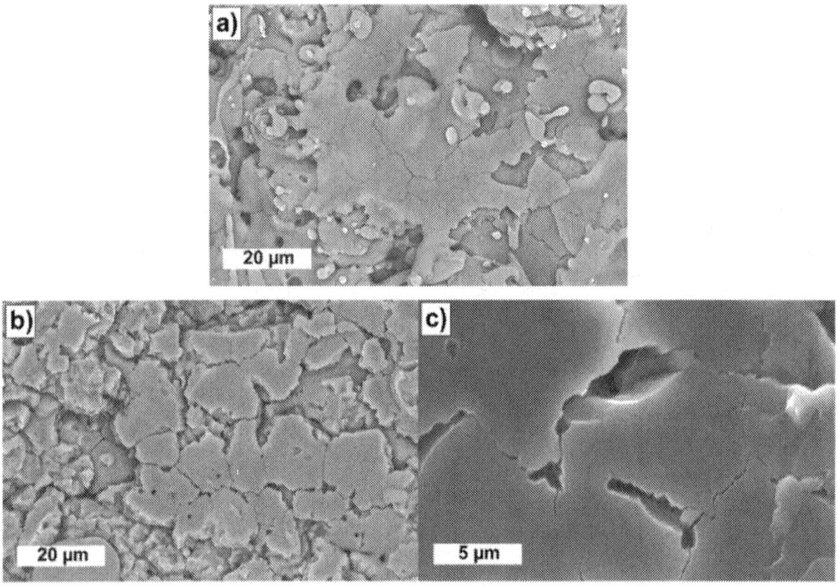

Figure 9. (**a**) Lamella on as-sprayed APS_A coating; (**b**) Lamella after cavitation erosion (APS_A); (**c**) Microcracks being worked at by cavitation erosion (APS_A).

When comparing APS- and HVOF-sprayed coatings and their wear surfaces, the wear mechanism seemed to be the same. As the particle size for HVOF-sprayed ceramic coatings are significantly smaller, also the lamellae size and thickness are smaller. This leads to smaller structural details of the coatings, as well as the wear surface. Another significant difference between plasma- and HVOF-sprayed coatings from fused and crushed powders, is the size of the horizontal cracks existing between splats. Better bonding of HVOF coatings promotes the good cavitation erosion resistance.

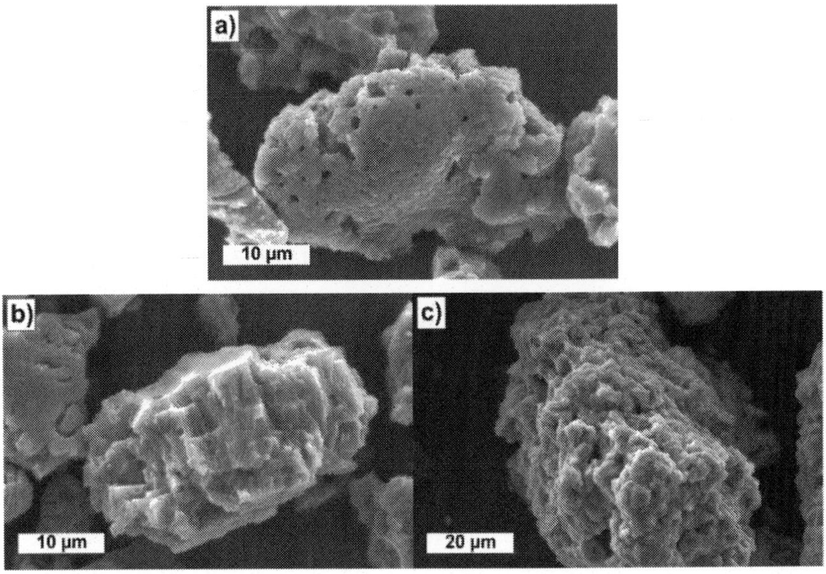

Figure 10. (**a**) Partially molten agglomerated and sintered Al_2O_3-13TiO_2 particle removed during cavitation erosion test (APS_AT2); (**b**) Well-melted region of the same coating removed during cavitation erosion test (APS_AT2); (**c**) Large debris particle from HVOF_AT2 coating.

3.4. Abrasion Wear Resistance

HVOF-sprayed ceramic coatings have generally been reported as having better mechanical properties and higher abrasion resistance [18,19] compared to their conventionally plasma-sprayed counterparts. The better performance is due to the smaller particle size and higher particle velocity during the spraying. These features of HVOF-sprayed ceramic coatings make it possible to produce dense coatings with fine microstructure. The weight losses after the rubber wheel dry-abrasion tests are shown in Figure 11. The high-abrasion-wear resistance of HVOF coatings sprayed from fused and crushed powders (HVOF A and HVOF_AT1) is clearly detected. Indeed, the performance of HVOF coatings was comparable with bulk Al_2O_3. Coatings sprayed from the agglomerated and sintered powders, on the other hand, did not rank in the same way, as the plasma-sprayed APS_AT2 coating performed better (86 mg) than the HVOF sprayed HVOF_AT2 coating (116 mg). Even though the HVOF sprayed coating seemed to be very dense in Figure 4f, the coating structure was found to contain more unmelted particles compared to the APS coating. This decreases the abrasion resistance in dry-abrasion rubber wheel test due to the fact that poorly adhered unmelted particles come loose more easily. Despite the insufficient melting of the particles, the HVOF_AT2 coating performed better than the APS-sprayed Al_2O_3 and Al_2O_3-TiO_2 (fused and crushed) coatings. The dense part of the HVOF_AT2 coating's structure appears to compensate the poor coating quality connected to unmelted particles.

When comparing the coatings sprayed from fused and crushed powders with APS and HVOF spray processes, the Al_2O_3coatings had better abrasion wear resistance than the coatings with 13% TiO_2 alloying. The TiO_2 addition lowers the hardness and abrasion resistance compared to pure Al_2O_3 [2]. Having a coating structure with varying layers of Al_2O_3 and TiO_2, e.g., coatings designated as APS_AT1 and HVOF_AT1, means that the wear resistance varies between splats of different material.

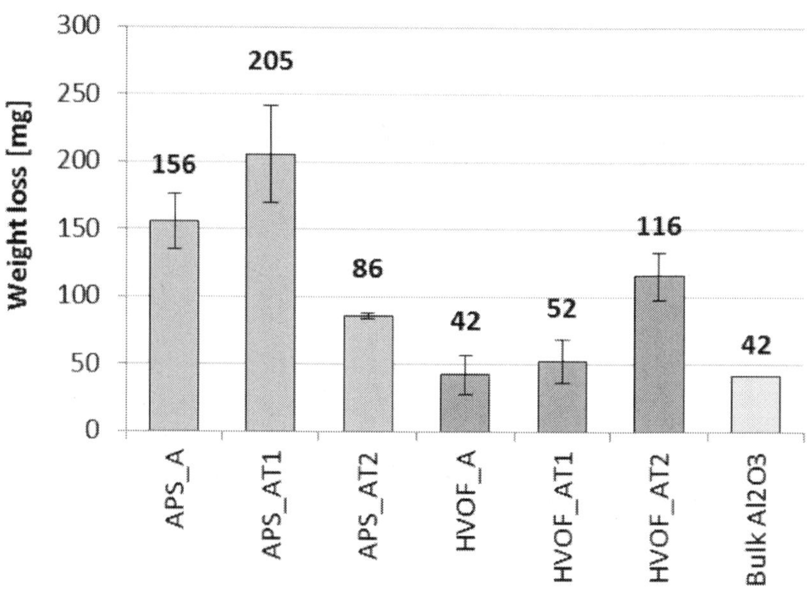

Figure 11. Weight losses and standard deviations of the APS and HVOF coatings and bulk Al_2O_3 material after abrasion wear test.

SEM studies revealed marks of ploughing and plastic deformation of the worn surfaces accompanied with areas of brittle fracture. The mechanism seemed to be the same for all surfaces, with significant differences only in the amount of brittle fracture areas on the surface arising from the lower ductility and cohesion of the coating, *i.e.*, tendency to brittle fracture. The APS_AT1 coating showed the lowest resistance to abrasive wear and the surface after the abrasion test is presented in Figure 12a. The surface consists of smoother plastically deformed areas and rough valleys showing signs of brittle fracture and unworn splat surfaces, which indicates that pieces of coating have been removed from the top. The wear surface of HVOF_A coating, which had the highest abrasive wear resistance, is shown in Figure 12b. It differs from the higher wear rate surface of APS_AT1 by having larger areas of plastically deformed smooth regions. The areas of brittle fracture seem shallower and smaller in size compared to APS_AT1. According to the abrasion wear results, the ranking of the coatings from highest wear resistance to lowest was HVOF_A, HVOF_AT1, APS_AT2, HVOF_AT2, APS_A, and APS_AT1.

Figure 12. Worn surfaces of (**a**) APS_AT1 and (**b**) HVOF_A coatings after 60-minute abrasion wear test.

3.5. Erosion Wear Resistance

Figure 13 shows the results (weight losses) from the centrifugal erosion wear tests at 30° and 90° impact angles. The erosion wear behavior of the coatings was similar to that of the abrasion wear behavior and the ranking of the coatings from best to worse was: HVOF_A, HVOF_AT1, APS_AT2, HVOF_AT2, APS_A, and APS_AT1. The samples fixed to smaller 30° impact angle eroded less during the test compared to 90° impact angle. Hard and brittle coatings perform well at low impact angles and reach their maximum wear rate at 90° angle [38]. The erosion rate of the 30° samples ranged between 25% and 50% of the 90° impact angle erosion rate.

In addition, in this case, worn surfaces were observed with SEM. The worn surfaces showed that the material is mostly removed by ploughing and brittle fracture at the low impact angles. At 90° impact angle, the material is plastically deformed in the impact and removed by brittle fracture. Similar findings were discovered by Westergård [39,40]. Figure 14a shows a single particle impact site on the fine ground surface of the HVOF_AT2 coating. The coating was plastically deformed and some brittle fracture can be seen on the impact mark edges. Parts of the erosive SiO_2 particle have fractured and attached on the impact site. This form of plastic deformation accompanied with minor brittle fracture of the coating is caused by smaller erosive particles with lower kinetic energy. The erosion mechanism with large erosive particles impacting at 90° is mostly brittle fracture of the lamellas. Pieces of fractured SiO_2 particles attached on the coating surface were observed at both impact angles. A thin layer of SiO_2, *i.e.*, tribofilm, was found on several plough-sites, at small impact angles. This can be seen in Figure 14b, where, also, debris from the erosive particle is left on the surface. Material around the wear mark has been removed by brittle fracture caused by particle impacts on the surface.

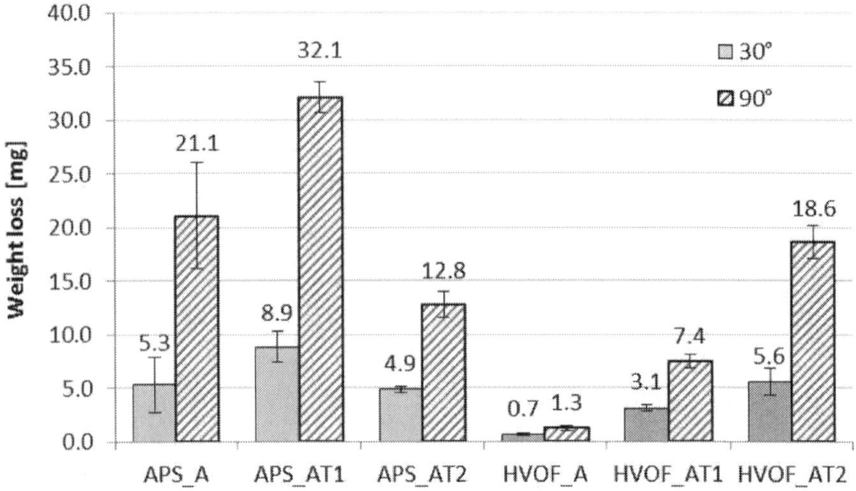

Figure 13. Weight losses and standard deviations of the APS (green) and HVOF (blue) coatings in the erosion wear test.

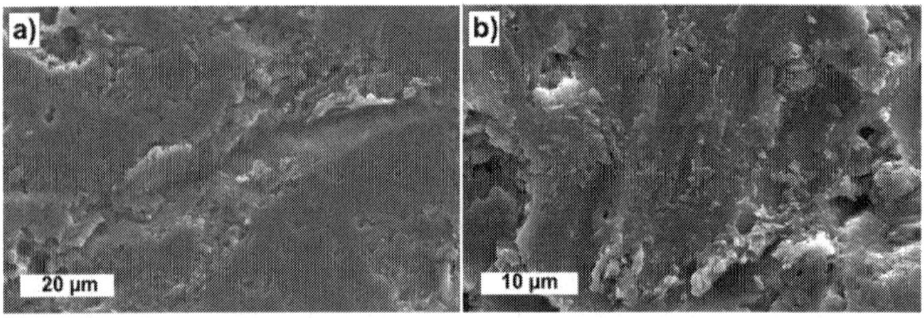

Figure 14. Worn surfaces of (**a**) HVOF_AT2 and (**b**) APS_AT2 coatings after the erosion test.

4. CONCLUSIONS

Abrasion, erosion, and cavitation erosion wear behavior of APS- and HVOF-sprayed Al_2O_3 and Al_2O_3-TiO_2 coatings sprayed with different powder morphologies were studied and compared. Following conclusions can be made from the results:

- Relatively dense coatings were sprayed from all powders with both spray processes. HVOF-sprayed coatings showed denser structure as higher particle velocities were reached in the process. This resulted in dense structure with small number of horizontal cracks. Larger cracks were observed in plasma-sprayed coatings compared to HVOF coatings.

- HVOF coatings, sprayed from fused and crushed powders, outperformed all other coatings in the wear tests. Higher velocity and small particle size resulted in dense and fine microstructure, which provided the coatings with high wear resistance comparable to bulk Al_2O_3.
- Plasma-sprayed coatings from agglomerated and sintered Al_2O_3-TiO_2 powder showed high wear resistance and outperformed the conventional coatings sprayed with fused and crushed powders. Improved properties were achieved by the agglomeration and sintering powder manufacturing route. This promoted the even distribution of TiO_2 in the coating.
- Coating ranking was basically identical between the abrasion and erosion wear tests as coating wear behavior was very similar. The denser HVOF coatings performed better than their APS counterparts. Improved abrasion and erosion resistance was achieved with the TiO_2 alloying as agglomerated and sintered powder was used. The TiO_2 addition, however, decreased the wear resistance when blended F/C powder was used.
- HVOF-sprayed coatings from F/C powders performed very well compared to other coatings. All the tested coatings showed linear weight loss. Material removal in the cavitation erosion test of the coatings happened purely by brittle fracture revealing the structural differences between thecoatings.

ACKNOWLEDGMENTS

The authors thank High-Alpha project, funded by industrial partners and the Finnish Funding Agency for Technology and Innovation (Tekes), for financial support of this work. Authors would also like to thank Mikko Kylmälahti of the Tampere University of Technology, Department of Materials Science, for spraying the coatings and Laura Gouillon (summer trainee at the Tampere University of Technology) for helping with the sample preparation and testing.

REFERENCES

1. Pawlowski, L. *The Science and Engineering of Thermal Spray Coatings*; John Wiley & Sons, Inc.: Oxford, UK, 1995; p. 626.
2. Davis, J.R. *Handbook of Thermal Spray Technology*; ASM International: Materials Park, OH, USA, 2004; p. 338.
3. Bengisu, M. *Engineering Ceramics*; Springer: New York, NY, USA, 2001; p. 620.

4. Gell, M.; Jordan, E.H.; Sohn, Y.H.; Goberman, D.; Shaw, L.; Xiao, T.D. Development and implementation of plasma sprayed nanostructured ceramic coatings. *Surf. Coatings Technol.* **2001**, *146–147*, 48–54.
5. Jordan, E.H.; Gell, M.; Sohn, Y.H.; Goberman, D.; Shaw, L.; Jiang, S.; Wang, M.; Xiao, T.D.; Wang, Y.; Strutt, P. Fabrication and evaluation of plasma sprayed nanostructured alumina-titania coatings with superior properties.*Mater. Sci. Eng. A* **2001**, *301*, 80–89.
6. Varis, T.; Knuuttila, J.; Turunen, E.; Leivo, J.; Silvonen, J.; Oksa, M. Improved protection properties by using nanostructured ceramic powders for HVOF coatings. *J. Therm. Spray Technol.* **2007**, *16*, 524–532.
7. Shaw, L.L.; Goberman, D.; Ren, R.; Gell, M.; Jiang, S.; Wang, Y.; Xiao, T.D.; Strutt, P.R. The dependency of microstructure and properties of nanostructured coatings on plasma spray conditions. *Surf. Coatings Technol.* **2000**,*130*, 1–8.
8. Ahn, J.; Hwang, B.; Song, E.P.; Lee, S.; Kim, N.J. Correlation of microstructure and wear resistance of Al_2O_3-TiO_2coatings plasma sprayed with nanopowders. *Met. Mater. Trans. A* **2006**, *37*, 1851–1861.
9. Wang, M.; Shaw, L.L. Effects of the powder manufacturing method on microstructure and wear performance of plasma sprayed alumina-titania coatings. *Surf. Coatings Technol.* **2007**, *202*, 34–44.
10. Lima, R.S.; Moreau, C.; Marple, B.R. HVOF-sprayed coatings engineered from mixtures of nanostructured and submicron Al_2O_3-TiO_2 powders: An enhanced wear performance. *J. Therm. Spray Technol.* **2007**, *16*, 866–872.
11. Fauchais, P. Understanding plasma spraying. *J. Phys. D Appl. Phys.* **2004**, *37*, R86–R108.
12. Niemi, K.; Vuoristo, P.; Mäntylä, T. Properties of Alumina Based Coatings Deposited by Plasma and High Velocity Combustion Processes. In Proceedings of 1993 National Thermal Spray Conference, Anaheim, CA, USA, 7–11 June 1993; pp. 469–474.
13. Niemi, K.; Vuoristo, P.; Mäntylä, T. Properties of alumina-based coatings deposited by plasma spray and detonation gun spray processes. *J. Therm. Spray Technol.* **1994**, *3*, 199–203.
14. Niemi, K. Abrasion Wear Characteristics of Thermally Sprayed Alumina Based Coatings. Ph.D. Thesis, Tampere University of Technology, Tampere, Finland, 2009; p. 117.
15. Niemi, K.; Hakalahti, J.; Hyvärinen, L.; Laurila, J.; Vuoristo, P.; Berger, L.-M.; Toma, F.-L.; Shahkhverdova, I. Influence of chromia alloying on the characteristics of APS and HVOF sprayed alumina coatings. In Proceedings of International Thermal Spray Conference 2011, Hamburg, Germany, 27–29 September 2011.
16. Toma, F.-L.; Scheitz, S.; Berger, L.-M.; Sauchuk, V.; Kusnezoff, M.; Thiele, S. Comparative study of electrical properties and characteristics of

thermally sprayed alumina and spinel coatings. *J. Therm. Spray Technol.* **2011**, *20*, 195–204.
17. Kreye, H.; Zimmermann, S.; Heinrich, P. The role of the fuel gas in the HVOF process. In Proceedings of ITSC'95, Kobe, Japan, 22–26 May 1995; ASM International: Materials Park, OH, USA, 1995; pp. 393–398.
18. Bolelli, G.; Lusvarghi, L.; Manfredini, T.; Mantini, F.P.; Polini, R.; Turunen, E.; Varis, T. Comparison between plasma- and HVOF-sprayed ceramic coatings. Part I: Microstructure and mechanical properties. *Int. J. Surf. Sci. Eng.* **2007**, *1*, 38–61.
19. Vuoristo, P.; Niemi, K.; Matikainen, V.; Hyvärinen, L.; Koivuluoto, H.; Berger, L.-M.; Scheitz, S.; Shakhverdova, I. Structure and Properties of HVOF and Plasma Sprayed Ceramic Alumina-Chromia Coatings Deposited from Fused and Crushed Powders. In Thermal Spray 2013—Innovative Coating Solutions for the Global Economy, Proceedings of the ITSC2013 Conference, Busan, Korea, 13–15 May 2013; Lima, R.S., Agarwal, A., Hyland, M.M., Lau, Y.-C., Mauer, G., McDonald, A., Toma, F.-L., Eds.; ASM International: Materials Park, OH, USA, 2013; pp. 465–470.
20. Singh, R.; Tiwari, S.K.; Mishra, S.K. Cavitation erosion in hydraulic turbine components and mitigation by coatings: Current status and future needs. *J. Mater. Eng. Perform.* **2011**, *21*, 1539–1551.
21. Sugiyama, K.; Nakahama, S.; Hattori, S.; Nakano, K. Slurry wear and cavitation erosion of thermal-sprayed cermets. *Wear* **2005**, *258*, 768–775.
22. Yuping, W.; Pinghua, L.; Chenglin, C.; Zehua, W.; Ming, C.; Junhua, H. Cavitation erosion characteristics of a Fe-Cr-Si-B-Mn coating fabricated by high velocity oxy-fuel (HVOF) thermal spray. *Mater. Lett.* **2007**, *61*, 1867–1872.
23. Lima, M.M.; Godoy, C.; Modenesi, P.J.; Avelar-Batista, J.C.; Davison, A.; Matthews, A. Coating fracture toughness determined by vickers identation: An important parameter in cavitation erosion resistance of WC-Co thermally sprayed coatings. *Surf. Coatings Technol.* **2004**, *177–178*, 489–496.
24. Tomlinson, W.J.; Kalitsounakis, N.; Vekinis, G. Cavitation erosion of aluminas. *Ceram. Int.* **1999**, *25*, 331–338.
25. Niebuhr, D. Cavitation erosion behavior of ceramics in aqueous solutions. *Wear* **2007**, *263*, 295–300.
26. Fatjó, G.G.; Hadfield, M.; Vieillard, C.; Sekulic, J. Early stage cavitation erosion within ceramics—An experimental investigation. *Ceram. Int.* **2009**, *35*, 3301–3312.
27. Jafarzadeh, K.; Valefi, Z.; Ghavidel, B. The effect of plasma spray parameters on the cavitation erosion of Al_2O_3-TiO_2coatings. *Surf. Coatings Technol.* **2010**, *205*, 1850–1855.

28. Schwetzke, R.; Kreye, H. Cavitation Erosion of HVOF Coatings. In *Thermal Spray: Practical Solutions for Engineering Problems*; Berndt, C.C., Ed.; ASM International: Materials Park, OH, USA, 1996; pp. 153–158.
29. Espitia, L.A.; Toro, A. Cavitation resistance, microstructure and surface topography of materials used for hydraulic components. *Tribol. Int.* **2010**, *43*, 2037–2045.
30. *ASTM G32-10 Standard Test Method for Cavitation Erosion Using Vibratory Apparatus*; ASTM International: West Conshohocken, PA, USA, 2010.
31. Kuroda, S.; Dendo, T.; Kitahara, S. Quenching stress in plasma sprayed coatings and its correlation with the deposit microstructure. *J. Therm. Spray Technol.* **1995**, *4*, 75–84.
32. Li, L.; Kharas, B.; Zhang, H.; Sampath, S. Suppression of crystallization during high velocity impact quenching of alumina droplets: Observations and characterization. *Mater. Sci. Eng. A* **2007**, *456*, 35–42.
33. Sakoda, T.; Inaba, M.; Taguchi, K. Microscopic Characteristics and Mechanical Properties of Plasma Sprayed Al_2O_3Coatings. In Thermal Spray 2013—Innovative Coating Solutions for the Global Economy, Proceedings of the ITSC2013 Conference, Busan, Korea, 13–15 May 2013; Lima, R.S., Agarwal, A., Hyland, M.M., Lau, Y.-C., Mauer, G., McDonald, A., Toma, F.-L., Eds.; ASM International: Materials Park, OH, USA, 2013; pp. 406–411.
34. Hurevich, V.; Smurov, I.; Pawlowski, L. Theoretical study of the powder behavior of porous particles in a flame during plasma spraying. *Surf. Coatings Technol.* **2002**, *151–152*, 370–376.
35. Tsochatzidis, N.A.; Guiraud, P.; Wilhelm, A.M.; Delmas, H. Determination of velocity, size and concentration of ultrasonic cavitation bubbles by the phase-Doppler technique. *Chem. Eng. Sci.* **2001**, *56*, 1831–1840.
36. Karimi, A.; Martin, J.L. Cavitation erosion of materials. *Int. Met. Rev.* **1986**, *31*, 1–26.
37. Tucker, R.C. *ASM Handbook Volume 5A: Thermal Spray Technology*; ASM International: Materials Park, OH, USA, 2013.
38. American Society for Metals. *ASM Handbook Volume 18: Friction, Lubrication, and Wear Technology*, 10th ed.; ASM International: Materials Park, OH, USA, 1992.
39. Westergård, R.; Erickson, L.C.; Axén, N.; Hawthorne, H.M.; Hogmark, S. The erosion and abrasion characteristics of alumina coatings plasma sprayed under different spraying coatings. *Tribol. Int.* **1998**, *31*, 271–279.
40. Westergård, R.; Axén, N.; Wiklund, U.; Hogmark, S. An evaluation of plasma sprayed ceramic coatings by erosion, abrasion and bend testing. *Wear* **2000**, *246*, 12–19.

Chapter 8

Titanium Nitride and Nitrogen Ion Implanted Coated Dental Materials

Youssef S. Al Jabbari[1], Jennifer Fehrman[2], Ashley C. Barnes[2], Angela M. Zapf[3], Spiros Zinelis[1,4] and David W. Berzins[1,3,]*

[1] *Dental Biomaterials Research and Development Chair, College of Dentistry, King Saud University, Riyadh 11545, Saudi Arabia; Email:*
[2] *School of Dentistry, Marquette University, Milwaukee, WI 53233, USA; Emails:*
[3] *Dental Biomaterials, School of Dentistry, Marquette University, Milwaukee, WI 53233, USA; Email:*
[4] *Department of Biomaterials, School of Dentistry, University of Athens, Athens, Greece*

ABSTRACT

Titanium nitride and/or nitrogen ion implanted coated dental materials have been investigated since the mid-1980s and considered in various applications in dentistry such as implants, abutments, orthodontic wires, endodontic files, periodontal/oral hygiene instruments, and casting alloys for fixed restorations. Multiple methodologies have been employed to create the coatings, but detailed structural analysis of the coatings is generally lacking in the dental literature. Depending on application, the purpose of the coating is to provide increased surface hardness, abrasion/wear resistance, esthetics, and corrosion resistance, lower friction, as well as greater beneficial interaction with adjacent biological and material substrates. While many studies have reported on the achievement of these properties, a consensus is not always clear. Additionally, few studies have been conducted to assess the efficacy of the coatings in a clinical setting.

Overall, titanium nitride and/or nitrogen ion implanted coated dental materials potentially offer advantages over uncoated counterparts, but more investigation is needed to document the structure of the coatings and their clinical effectiveness.

Keywords: titanium nitride; nitrogen ion implantation; dental materials; implants; endodontics; orthodontics; dental instruments

1. INTRODUCTION

Titanium nitride (TiN) and nitrogen ion implanted coatings have been used in industry for quite some time with eventual cross-over to dental applications in the mid-1980s. Since this time, these coatings have been applied to dental materials used in implant dentistry, orthodontics, endodontics, prosthodontics, and periodontics. Depending on application, the purpose of the coating is to provide increased surface hardness, abrasion/wear resistance, and corrosion resistance, lower friction, as well as greater beneficial interaction with adjacent biological and material substrates. A review of titanium nitride coatings in clinical dentistry was published in 1992 by Mezger and Creugers [1]. Taking into consideration that much has changed in dentistry and coating technology since 1992, the objective of this article was to provide a review of the literature of titanium nitride and nitrogen-ion implanted coated dental materials, focusing on coating methodologies, structural analysis, and an examination of coating applications and properties in the aforementioned disciplines.

2. COATING METHODOLOGIES AND STRUCTURAL ANALYSIS

A myriad of methodologies have been used to create TiN and nitrogen ion implanted coatings in various dental materials. A compilation of the methodologies as listed in the literature is displayed in Table 1. It is possible some subcategory methodologies are the same since the methodology to coat the dental materials was not always explicitly detailed or common nomenclature was not used. Still, it is apparent the majority of methodologies used are of the physical vapor deposition (PVD) type while thermal nitriding and chemical vapor deposition were employed in only 3 and 4 studies, respectively. Approximately 20 studies did not list the way in which the coating was deposited. A possible reason for this is that the coating may have been on a commercial product; thus the method/details of deposition may remain proprietary or are not published in manufacturer literature. Additionally, the coating was not the main focus of a few studies so details were understandably lacking. Coating methodology was contained in all of the Endodontic and Prosthodontic literature reviewed below as the coatings were more likely to be experimental in nature and certain properties were investigated, whereas it was less likely to be mentioned in the Dental Implant and Orthodontic literature; for the latter, commercial products were commonly evaluated.

Structural analysis, in the form of determining composition or phase structure of the coatings, was only used in a minority of studies (Table 1). Three methods were used to identify composition and/or phase structure: X-Ray Photoelectron Spectroscopy (XPS), X-Ray Diffraction (XRD), and Nuclear Reaction Analysis. XPS was used for detailed surface composition analysis. Often the N/Ti ratio was determined to indicate the amount of N deposition in the coating and to determine whether the ratio corresponded to stoichiometric values. Additionally, alternating XPS with sputter etching allowed depth profiles to be generated. XRD was used to identify the crystal structure of the coatings. A majority of studies that used XRD (5 of 8 in Table 1) identified only TiN, whereas TiN with Ti_2N was identified only once. Oxynitrides were also mentioned, but definitive identification was lacking. Nuclear Reaction Analysis was used in two studies to determine the concentration of nitrogen as a function of depth.

Table 1. Compilation of coating methods, structural (composition and phase) analysis methodology, and coating thickness listed in the various dental materials literature.

Coating Method	Type/Detail	Reference
Physical Vapor Deposition	No further detail beyond "Physical Vapor Deposition"	[2–7]
	Ion Plating	[8–14]
	Cathodic Arc	[15–19]
	Radiofrequency sputtering	[20–27]
	Arc ion plating	[28–31]
	Plasma immersion implantation	[32–34]
	Ion beam/Ionic implantation	[8,9,35–42]
Thermal	500 °C for 480 min	[35]
	850 °C for 7 h	[43]
	800–1000 °C for 5–30 min	[44,45]
Chemical Vapor/Chemical	Reaction of wet NH_3 with NiTi at 300 °C	[46]
	MOCVD (metal organic chemical vapor deposition) via evaporation of $Ti(Et_2N)_4$	[28,46]
	Plasma chemical vapor	[38]
	Boiling in 30% nitric acid for 24 h	[34]
No method stated		[47–68]

Structural Analysis	Comment	Reference
X-Ray Photoelectron Spectroscopy (XPS)	[N], [Ti], and sometimes other elements and ratios determined	[3,28,29, 32–35,46]
X-Ray Diffraction (XRD)	TiN Identified	[27,29,30, 42,67]
	TiN and Ti_2N Identified	[43]
	Unidentified oxynitrides	[44,45]
Nuclear Reaction Analysis	[N] depth	[36,37]

Method to Determine Coating Thickness	Thickness	Reference
Scanning Electron Microscopy (SEM)	1–2 μm	[29]
	2 μm	[20,43]
	~10 μm	[44,45]
Electron Probe MicroAnalysis (EPMA)	0.3 μm average	[26]

Method to Determine Coating Thickness	Thickness	Reference
XPS	$[N]_{max}$ within 10 nm; [N] detected up to 60 nm depth	[35]
Nuclear Reaction Analysis	[N] detected up to 60 nm	[36,37]
Ball-cratering method	1.5 μm	[21]
Light Microscopy	8 μm	[67]
Given via a reference, manufacturer data, equipment parameters, or other unconfirmed sources	<650 nm	[40]
	0.3 μm *N*-implanted; 3 μm TiN	[8,9]
	1 μm	[27]
	1.5 μm	[17,18]
	1.8 μm	[22–25]
	2 μm	[14,42]
	1–5 μm	[15]
	2–5 μm	[47]
	10 μm	[19]

Using these forms of structural analysis, it is apparent that the different coating methodologies will impart different deposition of nitrogen in terms of amount and depth, as also documented by others [28,32,35,46]. Even different parameters within a given coating technology will affect nitrogen distribution [36]. Overall, structural analysis of the coatings in the dental materials literature is lacking in a majority of studies.

The thickness of the TiN and nitrogen ion implanted coatings are also compiled in Table 1. The majority of studies list the coating thickness as between 1 μm and 5 μm although the range is anywhere from 0.06 μm to 10 μm. Scanning Electron Microscopy (SEM) was used most frequently to determine coating thickness; also employed were XPS and Nuclear Reaction Analysis as mentioned above, as well as Electron Probe MicroAnalysis (EPMA), light microscopy, and the ball-cratering method. However, it was most common to cite the thickness of coating via reference, manufacturing data, based upon equipment parameters, or other unspecified sources instead of actual measurement. Since the thickness of coatings produced by the methodologies listed in Table 1 are sensitive to various parameters (e.g., time and temperature for thermal nitridation), specifically measuring the coating thickness would seem to be the best practice.

Although the majority of analyses document more of a monolithic coating, it is likely the coatings exist as complex, layered structures. Certainly, the nitrogen concentration depth profiles may suggest this [28,32,35,36,37,46]. Additionally, several researchers observed multiple layers, whether intentionally deposited to increase adherence of the coating, a by-product of the coating process, or natural as in the form of titanium's passive oxide layer. For instance, Chung *et al.* [20] and Kurt *et al.* [21] used an intermediate layer of sputter coated Ti and Al before adding nitrogen to improve adhesion of layers on the substrate. On the other hand, Iijima *et al.* [33] observed TiO_2 on the surface covering a TiN layer on nickel-titanium (NiTi) wires via XPS, presumably due to the natural passivation of titanium. Similarly, Endo *et al.* [29] showed a trilayer composed of TiO_2 on the outside, a middle layer of TiN_x ($x > 1$), and an inner layer of TiN.

2.1. Coatings in Implant Dentistry

The clinical efficacy of dental implants depends on many different factors related to the bulk and surface properties of the dental implant materials and their components. Recently, the application of nitride coatings on Ti and its alloys have been proposed as a surface treatment aiming to increase the mechanical, physical, and esthetic properties of dental implants. Generally, nitride coatings, by a factor of approximately 10, increase the surface hardness [43,47], wear resistance [47], and corrosion resistance [29,43] of implant materials. Additionally beneficial, coated Ti surfaces have also been connected with antimicrobial properties [8,9]. Nitride coatings specific to dental implant materials have been prepared mainly by Physical Vapor Deposition methods, with the exception of Tamura *et al.* [43] who used a thermal treatment, with measured thicknesses ranging between 1 μm and 2 μm [21,29,43].

Bacterial adherence to the implant surface is associated with the pathogenesis of bacterial infections leading to implant loss. The implant abutment is exposed in the oral environment and so dental calculus and plaque can be readily developed. However, the calculus and plaque should be periodically removed to obtain a good prognosis throughout the lifetime of the dental implant. In light of this, antibacterial activity is a desirable property while, especially for the abutment, enhanced abrasion resistance against scaling treatments is wanted to maintain the surface finish of the abutment. However, surface coatings have been shown to be prone to detachment after just a few actions with Ti and stainless steel dental instruments [15].

Three *in vitro* research studies assessed the antibacterial activity of TiN coatings and concluded that there is no difference between Ti and nitrified Ti for adherence of *Streptococcus mutans* [43], *Porphyromonas gingivalis*, and *Actinobacillus actinomycetemcomitans* [8,9]. In contrast, the adherence of *Streptococcus mutans* and *Streptococcus sanguis* was significantly reduced on TiN surfaces compared to polished ones [22]. The same favorable results were found for TiN-TPS (titanium plasma sprayed) compared to uncoated TPS where *Streptococcus pyogenes* and *Streptococcus sanguinis* bacteria demonstrated decreased bacterial adhesion and proliferation [16]. Two studies based on *in vivo* data confirmed the positive effect of TiN coating on the antibacterial activity of Ti surfaces. The results of Scarano *et al.* [2] showed that implants coated with TiN illustrated a minor quantity of the surface covered by bacteria after a 24-h exposure to the oral cavity while a smaller bacterial quantity were found on TiN glass sheets compared to pure Ti coated glass sheets after 60 h of intraoral exposure [23]. TiN coatings seem to have a beneficial effect on antibacterial activity inhibiting the formation of microbial plaque, minimizing the adverse effects of peri-implantitis on implant longevity. The aforementioned antibacterial activity combined with the wear resistance to scaling treatment points out that the application of TiN coatings to implant abutments holds promise and is worthwhile for further scientific research and optimization.

The introduction of Ti and its alloys in implant dentistry was based on the osseointegration properties of Ti oxide (*i.e.*, anchoring of the implant via growth and attachment of bone to the implant). After this breakthrough, research worldwide focused on surface modifications to improve and accelerate bone remodeling on Ti surfaces. *In vitro* and *in vivo* results show that TiN coatings do not have an adverse effect on this vital property. *In vitro* studies demonstrated that cell adhesion was no different between coated and uncoated Ti surfaces employing either mouse [24] or human gingival fibroblasts [25], while proliferation of human osteoblasts increased on $TiNO_x$-coated titanium surfaces compared to uncoated controls [3]. Similar results were found by other studies applying bone marrow human mesenchymal stem cells [38] and bone marrow human stromal cells [4]. *In vivo* tests with animal models were similarly positive. Despite the diversity of experimental methodology, all *in vivo* data illustrate that bone formation as expressed by Bone Implant Contact (BIC) around coated and uncoated dental implants is similar. Table 2 demonstrates the duration of implantation, the animal model used, and BIC for Ti and TiN-coated materials from several *in vivo* studies. In yet another *in vivo* study in rats, the thickness of fibrous connective tissue around Ti-6Al-4V implants was found to be the same whether it was coated with TiO_2 or TiN [10]. Based on the above studies, it can be concluded that the TiN coating demonstrates a good biological response and does not negatively affect osseointegration and healing around Ti implants.

Table 2. Duration of implantation, animal model, and Bone Implant Contact (BIC) for several *in vivo* implant material studies.

Duration (Weeks)	Animal Model	Prior Surface Treatment	BIC (%)		Reference
			Ti	TiN	
8	Rats	Machining	59.0 ± 3.1	58.0 ± 3.0	[5]
8	Rats	Sandblasting	64.0 ± 4.1	66.0 ± 3.5	[5]
8	Rats	TPS	72.0 ± 2.5	70.0 ± 2.1	[5]
8	Rats		42.6	40.8	[43]
8	Dogs		69.7 ± 1.3	70.6 ± 1.7	[48]
12	Rabbits		58.8	52.7	[39]

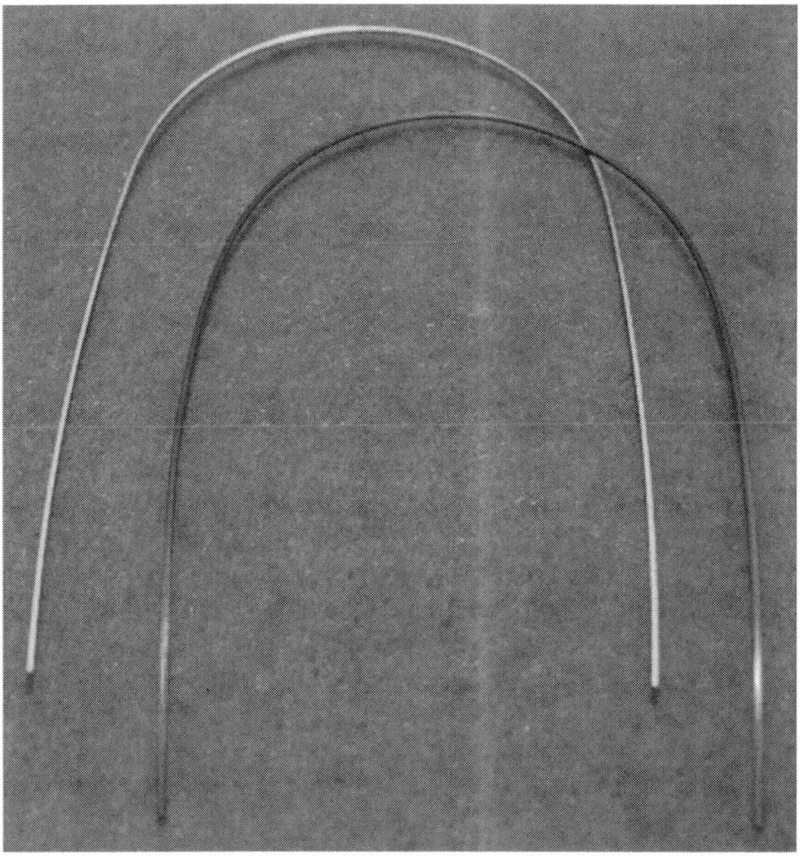

Figure 1. Ion implanted β-titanium orthodontic archwires (honeydew and purple).

Besides abutments and implants, TiN coatings have also been evaluated with other components. TiN coatings may provide enhanced resistance to wear in ball attachments for implant-overdenture systems, significantly increasing their longevity. After 1 year of clinical use, TiN coated attachment systems appeared unchanged while extensive wear was evident in uncoated systems [47]. Wear decreases the retention force of attachments, causing discomfort to the patient and increasing the maintenance needs of removable prosthetic restorations [49]. The last application of TiN coating in implant dentistry is the covering of retentions screws for implant supported crowns [50]. Screws coated with aluminum titanium nitride showed inferior retention of torque, a finding that might be associated with the minimal ductility of the coated screw compared to pure Ti. Thus, aluminum titanium nitride coatings are not recommended for this application.

In conclusion, the application of TiN coating on Ti implants provides antibacterial activity without compromising the osseointegrative properties of the Ti surface. In addition, it enhances the wear resistance of Ti abutments. Further research appears warranted, especially clinical research to assess the efficacy of the coatings long term.

2.2. Coatings in Orthodontics

During orthodontic treatment, tooth movement is achieved by applying forces to the teeth through the use of brackets, wires and elastics, which in turn result in the remodeling of the alveolus, or tooth socket. It is believed that light, continuous forces are optimal for most effectively and least traumatically moving teeth. The wire most commonly used in the initial leveling phase of treatment is the NiTi archwire, for it is the most elastic and flexible, capable of maintaining the light continuous forces desired. The NiTi alloy typically possesses a composition of 50/50 at % Ni/Ti, or approximately 55 wt % Ni and 45 wt % Ti, and was introduced to orthodontics in 1971 [69]. Treatment is typically finished out with β-titanium and stainless steel (SS) arch wires. β-titanium was introduced to orthodontics in 1979 [70] and contains Mo to metastabilize a BCC phase to room temperature, along with Sn and Zr. As a result, this wire was initially introduced by the Ormco Corporation (Glendora, CA, USA) as "TMA" for Titanium Molybdenum Alloy. While all three types of wires are available as originally designed, the NiTi and TMA wires are also available in versions which have undergone an ion implantation process during which nitrogen is implanted into the surface of the wires. The ion implanted TMA has also been known as "low-friction" and is offered in different colors as shown in Figure 1. This ion implantation process provides the NiTi and β-titanium wires with a coating which has been hypothesized to decrease surface roughness, and thereby decrease frictional forces, to produce more efficient tooth movement, and to reduce corrosion of the NiTi wires by limiting the release of nickel ions from the wire over time. Each of these hypotheses have been tested and documented in the literature. The following will provide an overview of the findings in regards to the effects of coatings applied to NiTi and β-titanium arch wires.

Table 3. Effect of nitrogen implantation on friction observed in several *in vitro*orthodontic studies.

Wire Type	Commercial Product Name (Manufacturer)	Coupled Material for Friction (if specified)	Amount of Friction Compared to Control (non-coated counterpart)	Reference
NiTi		Ti-6Al-4V	45%	[44]
NiTi		Stainless steel	59%	
NiTi	Neo Sentalloy F80 Ionguard (GAC Int., Bohemia, NY, USA)		77%	[51]
β-titanium	Purple TMA (Ormco)	Stainless steel	25% (static) 20% (kinetic)	[6]
β-titanium	Ion-implanted TMA (Ormco)	Ti implanted alumina	47% (static) 66% (kinetic)	[40]
β-titanium	Ion-implanted TMA (Ormco)	Stainless steel	100% (static) 108% (kinetic)	[54]
β-titanium	Aqua TMA (Ormco)	Stainless steel	105% (static) 111% (kinetic)	[55]
β-titanium	Honeydew TMA (Ormco)	Stainless steel	49% (static) 55% (kinetic)	
β-titanium	Purple TMA (Ormco)	Stainless steel	102% (static) 104% (kinetic)	
β-titanium	Violet TMA (Ormco)	Stainless steel	96% (static) 102% (kinetic)	
β-titanium	Ion-implanted TMA (Ormco)	Stainless steel	51% (static) 55% (kinetic)	

Perhaps the most widely tested hypothesis regarding ion implanted arch wires and/or brackets is whether these coatings have the ability to reduce surface roughness and frictional forces, and by extension, to produce more efficient tooth movement. Six studies [6,40,44,51,52,53] found that nitrogen implanted NiTi and TMA wires decreased surface roughness, frictional coefficients, or friction force values, although oral exposure may mitigate this initial advantage [51]. Contrastingly, two studies [11,54] found that applying coatings made no significant difference in the roughness and/or friction coefficients. Interestingly, another study found coated TMA exhibited increased or decreased static and kinetic friction forces compared to uncoated TMA depending on the color of the coating. This is curious since the TMA products are offered by the same company and presumably would originate from the same wire stock. Table

3 summarizes the effect of nitrogen implantation on friction behavior observed in some of the above orthodontic studies. When coatings did increase friction, the values were generally within 111% of that exhibited by the control or non-coated wire. However, when the coating decreased friction, values were as low as 20% of the control. Nonetheless, the stability of the coating *in vivo* and its ability to continue providing reduced friction is in question as shown by Wichelhaus *et al.* [51]. Although new wires of coated NiTi (Neo Sentalloy F80 Ionguard) displayed less friction compared to uncoated NiTi (Neo Sentalloy), after 4 weeks of clinical use, friction values were equivalent. Thus, while surface treatments were shown to improve the sliding of the bracket along the arch wire initially, that effect was lost when the wires were exposed to the oral cavity.

It has also been hypothesized that the ion implantation process would increase the biocompatibility of NiTi wires by reducing the amount of nickel released during the corrosion process. Seeing as nickel allergies are an issue in orthodontics, a reduction in the release of nickel ions would be greatly beneficial. Besides measuring nickel release into a solution, an indirect method of assessing this is to test the corrosion properties of the alloys. As a result, both methods appear in the literature. Of the studies which pertained to the above hypothesis, three [33,45,56] found a coating provided beneficial corrosion property alterations, while three other studies [12,57,58] found the ion implantation process incapable of decreasing the amount of nickel released into the oral environment or providing beneficial corrosion properties. These studies will be briefly summarized below.

NiTi wires treated with a nitrogen diffusion thermal treatment to create a TiN surface layer intended to reduce the corrosion potential of the wires were examined by Gil *et al.* [45] to measure ion release into an artificial saliva solution. While both the treated and untreated wires showed a high rate of nickel ion release initially, the treated wires ultimately released significantly less than that released by the untreated wires, reaching saturation at a much lower concentration of nickel ions. Iijima *et al.* [33] studied the corrosion properties of nitrogen ion implanted NiTi wires (Neo Sentalloy Ionguard) compared to non-ion implanted NiTi (Neo Sentalloy). The two wires were examined utilizing potentiodynamic polarization measurements, which were completed in both artificial saliva and a fluoride mouth rinse solution, as it has been hypothesized that the acidic environment created by fluoride may dissolve the TiN coating. Ultimately, it was found that the Neo Sentalloy Ionguard wires had a greater corrosion resistance in both artificial saliva and fluoride mouth rinse solution, despite the slightly more acidic environment that it created. Neumann *et al.* [56] similarly evaluated nitrogen ion implanted NiTi compared to an uncoated version, as well as two coated TMA wires, via measurement of rupture potentials in artificial saliva. The coating on the TMA wires was stable electrochemically; the coating on NiTi increased its rupture potential by 140 mV and reduced surface destruction when viewed under SEM.

Peitsch *et al.* [57] measured nickel release from mechanically loaded coated (Neo Sentalloy Ionguard) and uncoated NiTi in ultrapure water and saline. When the wires were subjected to a load, they released significantly more nickel ions, which is important to consider given that arch wires are typically under load in the oral cavity. However, the data could not confirm that surface nitridation reduced the release of nickel ions, whether in a loaded or unloaded state.

Kao *et al.* [12] conducted a study in which TiN ion implanted stainless steel brackets were analyzed against non-ion implanted stainless steel brackets in regards to corrosion resistance and release of Ni ions into the oral environment. The TiN coated brackets were shown to release more nickel, chromium, cobalt, manganese, and ferric ions. Copper was the only ion in which the ion implanted brackets released less of than the non-ion implanted brackets. Based on these findings, the authors were able to conclude that the anticorrosion properties of TiN coated brackets are relatively poor; thus no increase in biocompatibility is to be expected.

Kim *et al.* [58] studied the breakdown potential of the following arch wires: stainless steel, NiTi, nitride-coated NiTi, epoxy-coated NiTi, and titanium in a 0.9% NaCl solution of neutral pH at room temperature. It was determined that the TiN coating on the NiTi wires did not decrease corrosion and did not alter the breakdown potential. The epoxy-coated wire, however, did significantly decrease corrosion. Thus, the authors concluded that for patients with nickel allergies, an epoxy-coated NiTi wire would be recommended so as to reduce the risk of an allergic reaction.

Schuster *et al.* [59] determined the allergic potential of orthodontic materials utilized during treatment via questionnaire. While they concluded that the number of patients with allergies to orthodontic materials is likely overestimated, there are still 0.2% of patients who will show some sort of reaction. Based on the fact that allergies are a possibility, they recommend avoiding materials prone to corrosion, as well as coatings such as titanium nitride, so as not to induce a nickel related allergy. The rationale for the latter recommendation, however, was not well developed.

One additional study, conducted by Pernier *et al.* [60], analyzed the effects of autoclaving on ion implanted and non-ion implanted wires in regards to the effect of surface structure. Six commonly used wires were analyzed: one stainless steel wire, two NiTi wires (Neo Sentalloy and Neo Sentalloy Ionguard), and three titanium-molybdenum wires (TMA, Low Friction TMA and Resolve). The Low Friction TMA and NeoSentalloy Ionguard are commercially subjected to the ion implantation process. All wires were autoclaved for 18 min at 134 °C and examined before and after sterilization with various surface analysis techniques. None of the analyses showed significant changes in the properties of any of the wires after having been autoclaved. Thus, the authors concluded that autoclaving is a safe means of sterilizing arch wires without interfering with or changing the properties of such wires.

Lastly, it has been postulated that coated arch wires have the ability to generate more efficient tooth movement. Likely this finding is an extension of the effect the coatings have on the friction generated between the arch wire and the bracket. Of the studies which specifically analyzed efficiency of tooth movement, one [61] found that wires subjected to the nitrogen ion implantation process produced more efficient tooth movement, while two other studies [62,63] found that the ion implanted arch wires were not able to produce a significantly faster rate of tooth movement.

The following five wires were used by Ryan *et al.* [61] to test efficiency of tooth movement: N ion implanted nickel-titanium, untreated nickel-titanium, N ion implanted β-titanium, untreated β-titanium, and stainless steel. All wires were tested under identical conditions *in vitro*, and the relative tooth movement produced by each type of wire was measured and compared. The results showed that ion implantation of the wires was capable of reducing friction, and allowed the treated wires to produce significantly more tooth movement than their untreated counterparts.

In one of only two clinical studies looking at efficacy, Cobb *et al.* [62] measured the efficiency of initial alignment in 155 dental arches, with the patients being treated with stainless steel, superelastic NiTi, or ion implanted NiTi archwires. A three-way archwire randomization was completed for optimum results. The ultimate goal was to determine if the ion implantation process had the ability to produce more efficient tooth movement. Patients who were initially determined to have >5 mm incisor irregularity participated in this study over the course of 12 months. The degree of anterior irregularity was determined each month for each patient utilizing Little's irregularity index. Ultimately, it was determined that all three wires provided similar initial alignment when re-ligated on a monthly basis, showing no advantage to the ion implantation process for NiTi wires in terms of efficiency of tooth movement.

Kula *et al.* [63] conducted a "split mouth" design study in which 30 subjects were treated with β-titanium (TMA) wires, half of which had been ion implanted with nitrogen, the other half remaining in its natural state. At each monthly adjustment appointment, measurements were made to determine the amount of tooth movement achieved on each side of the mouth. Ultimately, they found that there was no statistically significant difference in the rates of closure between the implanted and unimplanted sides, thereby concluding that implanted TMA wires cannot enhance space closure.

In conclusion, while ion implanted arch wires have continually been gaining popularity due to their hypothesized reduction in friction, increase in rate of tooth movement and decrease in release of nickel ions, results are often contradictory and thus inconclusive. More clinical testing should be performed to determine with confidence whether surface-treated arch wires have the hypothesized benefits.

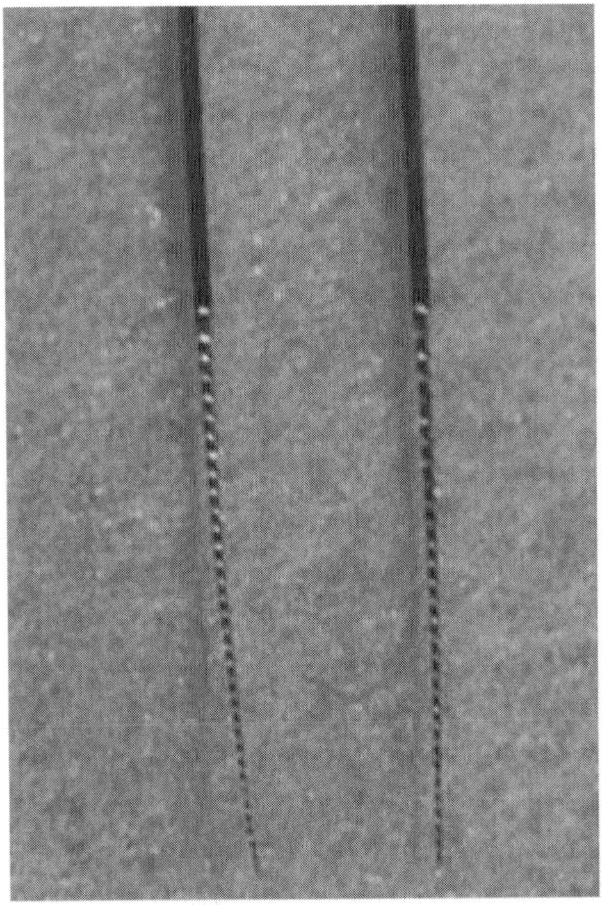

Figure 2. Titanium nitride coated (gold color; left) and uncoated endodontic files.

2.3. Coatings in Endodontics

Endodontic treatment involves the removal of infected nerve tissue within the pulp chamber and canals of the tooth. Due to the small size of the canals, endodontic files are used to shape and clean the canals along with specific chemicals to ensure the disinfection of the canals. Following its introduction to orthodontics, the benefits of NiTi in endodontic applications was first published in 1988 [71]. NiTi's superelastic characteristic allows for greater flexibility and elasticity to that of an ordinary metal/alloy; as a result it has superior ability compared to stainless steel to negotiate root canals without causing undesirable shapes where bacteria may reside. Although NiTi endodontic files show some ideal qualities, after repeated use, increased wear, lower cutting efficiency and reaching a cyclic fatigue limit can be problematic for the practitioner. Surface treatments, including TiN and N ion implantation, have been developed to help control these factors.Figure 2 displays a file that has a titanium nitride coating compared to an uncoated file.

As with other applications, some of the purported advantages associated with surface treatment include increased surface hardness, cutting efficiency, corrosion resistance, and cyclic fatigue resistance. Several studies have supported these statements. Rapisarda *et al.* [41] found both nitrogen ion implanted and thermally nitride files possessed greater wear resistance and had increased cutting capacity compared to uncoated controls. In a more qualitative study via SEM by the same authors [35], they found nitrogen ion implanted instruments did not manifest the typical signs of wear and surface changes that affected untreated NiTi files when used to instrument acrylic endodontic training blocks. Similarly, Schäfer [17] showed files that had undergone physical vapor deposition had increased cutting efficiency compared to uncoated files. Further, repeated sterilization cycles via autoclaving or exposure to sodium hypochlorite prior to sterilization did not alter the cutting efficiency of PVD-coated NiTi files but it did affect the non-coated files [18]. In this study, the greater cutting efficiency of PVD-coated files was noted among the non-sterilized groups. Bonaccorso *et al.* [7] observed greater pitting potentials of PVD-coated files tested in NaCl, indicating increased corrosion resistance. Gavini *et al.* [36] observed greater cyclic fatigue lifetimes in NiTi files processed via nitrogen ion implantation.

A potential issue when NiTi is subjected to coating treatments is whether the coating procedure will alter the phase transformation behavior of the files. For this reason, low temperature or even room temperature coating procedures have been advocated and shown to minimally affect phase transformation. Li *et al.* [34] found phase transformation temperatures and enthalpies between as-received and room temperature plasma immersion ion implantation files were similar, indicating the procedure should not affect superelastic properties.

Not all articles have provided positive results with coating endodontic files. Wolle *et al.* [37] compared Ni and Ar ion implanted files for fatigue resistance using curved canals in resin blocks. Results showed that argon implantation moderately improved the performance of the files, but nitrogen implanted files performed worse than uncoated files in the fatigue test. It was theorized that the reduction in file performance was caused by nitrogen diffusion in the grain boundaries, instead of the desired improvement caused by titanium nitride formation. Alves-Claro *et al.* [32] observed a decrease in hardness for NiTi files following plasma immersion nitrogen ion implantation, although they stated that wear resistance was increased.

Overall, the majority of articles presented suggest TiN coating of endodontic files leads to an increase in beneficial properties without necessarily sacrificing its superelastic qualities. However, it should be noted that these studies have all been *in vitro*, therefore the clinical success of coated NiTi endodontic instruments remains untested.

2.4. Coatings in Prosthodontics

Casting alloys used in dentistry may be bonded to a ceramic (porcelain) system for esthetic restorative purposes. In these systems, the bond between the ceramic material and the alloy relies on the oxide layer between the materials. Careful consideration of the material and methods to improve the oxide layer is required in order to increase the bond strength between the substrates. Although it has its drawbacks, titanium may be used in cast structures in dentistry due to its favorable properties of high strength, good biocompatibility, and corrosion resistance. When looking at titanium bonded to porcelain, the oxide layer can be controlled through the use of surface coatings. TiN coatings have been suggested to control the oxide layer formation and provide a better bond between materials. Oshida and Hashem [30] determined the oxide layer formed on samples coated with TiN resulted in a layer 2.34 times thinner compared to those samples without the coating. This was illustrated as the non-coated samples resulted in an oxide layer of 1–2 μm, whereas the TiN coated samples produced an oxide layer of 0.3–0.5 μm. These results indicate that the application of TiN as a coating does control the oxide layer. Whether this control over the oxide layer provided better bond strengths was not determined in this study.

Lim *et al.* [31] measured the fracture load of crowns using titanium copings and a low fusing porcelain to determine if a TiN coating provides better strength. In addition to a TiN coating, airborne particle abrasion with Al_2O_3 and sputter coating with gold were used for comparative surface preparation techniques. Of these, the gold coated titanium, as well as the TiN coated titanium copings, resulted in higher fracture loads than the airborne particle abraded group. This study demonstrated the application of TiN as a coating to titanium does provide strength similar to that of gold-ceramic crowns. Apart from porcelain adherence and strength, fit of a coping or crown is also important. Wu *et al.* [19] compared the marginal adaptation of a NiCr base metal alloy with and without a coating via optical microscopy. The TiN coated titanium copings resulted in clinically acceptable marginal adaption 64.86% of the time, whereas the base metal coping had an acceptance rate of only 47.50%.

Dental magnetic attachments have been used to aid stabilization of removable partial dentures and retention of implant-supported structures. Likewise, TiN coatings have been studied to determine the benefit of application on magnetic stainless steel attachment systems. Traditionally, magnetic stainless steel attachment systems have an undesirable level of corrosion, especially when coupled with other alloys in the oral cavity. The purpose of introducing a TiN coating on magnetic stainless steel has been to reduce corrosion. Hai *et al.* [13] compared the corrosion rates of a TiN coated magnetic stainless steel attachment system to pure titanium and found it did decrease the level of corrosion. While this study showed a decrease in corrosion of magnetic stainless steel upon application of TiN as a coating, the adhesive capabilities with other materials should also be considered. Taira *et al.* [14] determined the shear bond strengths of TiN coated magnetic stainless steel and non-coated magnetic stainless steel to two primers and bonding agents and found no significant difference in bond strength between any of the groups and across the materials (coated and non-coated). Therefore, due to the increase in corrosion resistance and equivalent bonding to other substrates, it may be suggested that TiN coated magnetic stainless steel should be used over non-coated magnetic stainless steel attachments. Titanium nitride coatings have since been further investigated in bonding with resin composites when applied to Au-Pd-Ag alloys, as in the case of a resin-veneered crown. Tanaka *et al.* [26] examined the shear bond strength of TiN coatings on Au-Pd-Ag alloys when bonded to composite resin with a metal primer. The results from this study showed an increase in bond strength for groups with a TiN coating on the alloy compared to the non-coated group.

Throughout the TiN coating prosthodontics literature, most research has demonstrated benefits for application of the coating. These benefits range from increases in physical properties to decreases in the amount of corrosion. Biocompatibility of the TiN coating to its oral surroundings is also very important clinically. Chien *et al.* [27] evaluated the cytotoxicity of nickel-based alloy surfaces after nitride film coatings of TiN and TiAlN. At 3 h, human fibroblast cells began to spread across all test surfaces, but at 24 h, the coated groups were significantly higher in cell proliferation and viability than polished and control Ni-alloy surfaces. Thus, the biocompatibility of this nickel-based alloy increased significantly after the application of TiN and TiAlN coatings.

Better marginal adaptation was found clinically with the application of TiN coating on titanium. For magnetic stainless steel alloys, better corrosion resistance and adequate bond strengths were found with the application of TiN as a coating. In addition, TiN coatings on Au-Pd-Ag alloys did increase the bond strength to composite resin and the coating of TiN to a nickel-based alloy improved biocompatibility. Overall, with all of these benefits in mind, it is clear that many dental alloys have marked improvements with the application of TiN coatings.

2.5. Coatings in Dental Instrumentation

Periodontal therapy includes the removal of hard and soft deposits from root surfaces. As a result, many dental instruments used in oral hygiene control must be sufficiently hard and resistant to wear and dulling. Naturally, coating dental instruments with titanium nitride coatings has been investigated as a means of developing instruments with better edge retention. Sisera *et al.* [64] compared TiN coated curettes to a cryogenically treated stainless steel instrument and uncoated stainless steel control in a study involving root planing of bovine teeth. All three materials held their cutting capacity for up to 1010 strokes, however, sterilization of the instruments was found to decrease cutting edge retention. Clinically, the cutting efficacy of the curettes is important for removal of subgingival biofilm, calculus, and endotoxins from the root surface. Aspriello *et al.* [65] showed the use of a curette followed by a TiN coated ultrasonic instrument created root surfaces free from smear layers, which presumably would promote early wound healing. Sawase *et al.* [42] measured surface hardness and abrasion resistance of different hygiene instruments, some with a TiN coating. The TiN coated titanium possessed a hardness approximately twelve times that of non-coated titanium. Further, abrasion resistance was greater in the TiN coated group compared to pure titanium.

Certain dental instruments are used to place restorative materials and may be used mechanically but may also contact restorative materials. Thus, titanium nitride has been researched as a coating to stainless steel dental instruments to improve properties of adhesion of the restorative material, contamination and discoloration of the restorative material, and wear resistance [66]. With regard to adhesion of restorative materials, TiN coatings have been suggested and marketed to be low-stick and improve this quality over traditional stainless steel instruments due to its low coefficient of friction and higher contact angle. For composite resins and glass ionomers, TiN coated stainless steel instruments produced no decrease in adhesion compared to non-coated, clean, polished stainless steel instruments. When considering discoloration of the restorative material, TiN coated stainless steel instruments produced no discoloration of the restoration, whereas non-coated stainless steel instruments did discolor composite resin samples [66]. When compared with non-coated stainless steel instruments, TiN coated instruments had better wear resistance to unpolymerized composite resin [67]. Further, this coating was shown to be harder, smoother and uncracked compared to alumina coated stainless steel instruments and non-coated stainless steel instruments after sterilization and ultrasonic cleaning cycles. In order to improve wear resistance of tungsten carbide (WC-Co) instruments, TiN has been suggested as an interlayer prior to diamond deposition on these dental instruments. In this case, TiN acts as a diffusion barrier between cobalt and carbon species to increase the adhesion of diamond films which are desired for their wear resistance [68]. When compared to tungsten carbide instruments without an interlayer of TiN, the instruments with TiN present increased the lifetime of the instrument and had better wear resistance [68].

3. CONCLUSIONS

Titanium nitride and/or nitrogen ion implanted coated dental materials have been investigated within many different dental disciplines for a few decades. Despite this, commercially-available materials primarily consist of coated orthodontic archwires. Though much research has determined the effect of the coatings on hardness, wear resistance, corrosion, friction, bone adaption, and bacterial adherence, very few clinical studies are available to ascertain their effectiveness *in vivo*.

ACKNOWLEDGMENTS

The authors thank Neill H. Luebke and the Ormco Corporation for providing the coated endodontic file and orthodontic wires, respectively, displayed in the figures.

REFERENCES

1. Mezger, P.R.; Creugers, N.H. Titanium nitride coatings in clinical dentistry. *J. Dent.* **1992**, *20*, 342–344.
2. Scarano, A.; Piattelli, M.; Vrespa, G.; Caputi, S.; Piattelli, A. Bacterial adhesion on titanium nitride-coated and uncoated implants: An *in vivo* human study. *J. Oral Implant.* **2003**, *29*, 80–85.
3. Durual, S.; Pernet, F.; Rieder, P.; Mekki, M.; Cattani-Lorente, M.; Wiskott, H.W. Titanium nitride oxide coating on rough titanium stimulates the proliferation of human primary osteoblasts. *Clin. Oral Implant. Res.* **2011**, *22*, 552–559.
4. Annunziata, M.; Guida, L.; Perillo, L.; Aversa, R.; Passaro, I.; Oliva, A. Biological response of human bone marrow stromal cells to sandblasted titanium nitride-coated implant surfaces. *J. Mater. Sci. Mater. Med.* **2008**, *19*, 3585–3591.
5. Scarano, A.; Piattelli, M.; Vrespa, G.; Petrone, G.; Iezzi, G.; Piattelli, A. Bone healing around titanium and titanium nitride-coated dental implants with three surfaces: An experimental study in rats. *Clin. Implant Dent. Relat. Res.* **2003**,*5*, 103–111.
6. Burstone, C.J.; Farzin-Nia, F. Production of low-friction and colored TMA by ion implantation. *J. Clin. Orthod.* **1995**,*29*, 453–461.
7. Bonaccorso, A.; Tripi, T.R.; Rondelli, G.; Condorelli, G.G.; Cantatore, G.; Schäfer, E. Pitting corrosion resistance of nickel-titanium rotary instruments with different surface treatments in seventeen percent

ethylenediaminetetraacetic acid and sodium chloride solutions. *J. Endod.* **2008**, *34*, 208–211.
8. Yoshinari, M.; Oda, Y.; Kato, T.; Okuda, K. Influence of surface modifications to titanium on antibacterial activity *in vitro*. *Biomaterials* **2001**, *22*, 2043–2048.
9. Yoshinari, M.; Oda, Y.; Kato, T.; Okuda, K.; Hirayama, A. Influence of surface modifications to titanium on oral bacterial adhesion *in vitro*. *J. Biomed. Mater. Res.* **2000**, *2*, 388–394.
10. Satomi, K.; Akagawa, Y.; Nikai, H.; Tsuru, H. Tissue response to implanted ceramic-coated titanium alloys in rats. *J. Oral Rehabil.* **1988**, *15*, 339–345.
11. Kao, C.T.; Guo, J.U.; Huang, T.H. Comparison of friction force between corroded and noncorroded titanium nitride plating of metal brackets. *Am. J. Orthod. Dentofac. Orthoped.* **2011**, *139*, 594–600.
12. Kao, C.T.; Ding, S.J.; Chen, Y.C.; Huang, T.H. The anticorrosion ability of titanium nitride (TiN) plating on an orthodontic metal bracket and its biocompatibility. *J. Biomed. Mater. Res.* **2002**, *63*, 786–792.
13. Hai, K.; Sawase, T.; Matsumura, H.; Atsuta, M.; Baba, K.; Hatada, R. Corrosion resistance of a magnetic stainless steel ion-plated with titanium nitride. *J. Oral Rehabil.* **2000**, *27*, 361–366.
14. Taira, Y.; Hai, K.; Matsumura, H.; Atsuta, M. Adhesive bonding of titanium nitride-plated stainless steel for magnetic attachments. *Eur. J. Oral Sci.* **2001**, *109*, 204–207.
15. Mengel, R.; Meer, C.; Flores-de-Jacoby, L. The treatment of uncoated and titanium nitride-coated abutments with different instruments. *Int. J. Oral Maxillofac. Implant.* **2004**, *19*, 232–238.
16. Annunziata, M.; Oliva, A.; Basile, M.A.; Giordano, M.; Mazzola, N.; Rizzo, A.; Lanza, A.; Guida, L. The effects of titanium nitride-coating on the topographic and biological features of TPS implant surfaces. *J. Dent.* **2011**, *39*, 720–728.
17. Schäfer, E. Effect of physical vapor deposition on cutting efficiency of nickel-titanium files. *J. Endod.* **2002**, *28*, 800–802.
18. Schäfer, E. Effect of sterilization on the cutting efficiency of PVD-coated nickel-titanium endodontic instruments. *Int. Endod. J.* **2002**, *35*, 867–872.
19. Wu, J.C.; Lai, L.C.; Sheets, C.G.; Earthman, J.; Newcomb, R. A comparison of the marginal adaptation of cathode-arc vapor-deposited titanium and cast base metal copings. *J. Prosthet. Dent.* **2011**, *105*, 403–409.
20. Chung, K.H.; Duh, J.G.; Shin, D.; Cagna, D.R.; Cronin, R.J., Jr. Characteristics and porcelain bond strength of (Ti,Al)N coating on dental alloys. *J. Biomed. Mater. Res.* **2002**, *63*, 516–521.

21. Kurt, M.; Külünk, T.; Ural, C.; Külünk, S.; Danişman, S.; Savaş, S. The effect of different surface treatments on retention of cement retained implant supported restorations. *J. Oral Implant.* **2010**.
22. Grössner-Schreiber, B.; Griepentrog, M.; Haustein, I.; Müller, W.D.; Lange, K.P.; Briedigkeit, H.; Göbel, U.B. Plaque formation on surface modified dental implants: An *in vitro* study. *Clin. Oral Implant. Res.* **2001**, *12*, 543–551.
23. Groessner-Schreiber, B.; Hannig, M.; Dück, A.; Griepentrog, M.; Wenderoth, D.F. Do different implant surfaces exposed in the oral cavity of humans show different biofilm compositions and activities? *Eur. J. Oral Sci.* **2004**, *112*, 516–522.
24. Groessner-Schreiber, B.; Neubert, A.; Müller, W.D.; Hopp, M.; Griepentrog, M.; Lange, K.P. Fibroblast growth on surface-modified dental implants: An *in vitro* study. *J. Biomed. Mater. Res. A* **2003**, *64*, 591–599.
25. Grössner-Schreiber, B.; Herzog, M.; Hedderich, J.; Dück, A.; Hannig, M.; Griepentrog, M. Focal adhesion contact formation by fibroblasts cultured on surface-modified dental implants: An *in vitro* study. *Clin. Oral Implant. Res.* **2006**,*17*, 736–745.
26. Tanaka, K.; Kimoto, K.; Sawada, T.; Toyoda, M. Shear bond strength of veneering composite resin to titanium nitride coating alloy deposited by radiofrequency sputtering. *J. Dent.* **2006**, *34*, 277–282.
27. Chien, C.C.; Liu, K.T.; Duh, J.G.; Chang, K.W.; Chung, K.H. Effect of nitride film coatings on cell compatibility. *Dent. Mater.* **2008**, *24*, 986–993.
28. Tripi, T.R.; Bonaccorso, A.; Rapisarda, E.; Tripi, V.; Condorelli, G.G.; Marino, R.; Fragalà, I. Depositions of nitrogen on NiTi instruments. *J. Endod.* **2002**, *28*, 497–500.
29. Endo, K.; Sachdeva, R.; Araki, Y.; Ohno, H. Effects of titanium nitride coatings on surface and corrosion characteristics of Ni-Ti alloy. *Dent. Mater. J.* **1994**, *13*, 228–239.
30. Oshida, Y.; Hashem, A. Titanium-porcelain system Part I: Oxidation kinetics of nitrided pure titanium, simulated to porcelain firing process. *Biomed. Mater. Eng.* **1993**, *3*, 185–198.
31. Lim, H.P.; Kim, J.H.; Lee, K.M.; Park, S.W. Fracture load of titanium crowns coated with gold or titanium nitride and bonded to low-fusing porcelain. *J. Prosthet. Dent.* **2011**, *10*, 164–170.
32. Alves-Claro, A.P.; Claro, F.A.; Uzumaki, E.T. Wear resistance of nickel-titanium endodontic files after surface treatment. *J. Mater. Sci. Mater. Med.* **2008**, *19*, 3273–3277.
33. Iijima, M.; Yuasa, T.; Endo, K.; Muguruma, T.; Ohno, H.; Mizoguchi, I. Corrosion behavior of ion implanted nickel-titanium orthodontic wire in fluoride mouth rinse solutions. *Dent. Mater. J.* **2010**, *29*, 53–58.

34. Li, U.M.; Iijima, M.; Endo, K.; Brantley, W.A.; Alapati, S.B.; Lin, C.P. Application of plasma immersion ion implantation for surface modification of nickel-titanium rotary instruments. *Dent. Mater. J.* **2007**, *26*, 467–473.
35. Rapisarda, E.; Bonaccorso, A.; Tripi, T.R.; Fragalk, I.; Condorelli, G.G. The effect of surface treatments of nickel-titanium files on wear and cutting efficiency. *Oral Surg. Oral Med. Oral Pathol. Oral Radiol. Endod.* **2000**, *89*, 363–368.
36. Gavini, G.; Pessoa, O.F.; Barletta, F.B.; Vasconcellos, M.A.; Caldeira, C.L. Cyclic fatigue resistance of rotary nickel-titanium instruments submitted to nitrogen ion implantation. *J. Endod.* **2010**, *36*, 1183–1186.
37. Wolle, C.F.; Vasconcellos, M.A.; Hinrichs, R.; Becker, A.N.; Barletta, F.B. The effect of argon and nitrogen ion implantation on nickel-titanium rotary instruments. *J. Endod.* **2009**, *35*, 1558–1562.
38. Catledge, S.A.; Vohra, Y.K.; Bellis, S.L.; Sawyer, A.A. Mesenchymal stem cell adhesion and spreading on nanostructured biomaterials. *J. Nanosci. Nanotechnol.* **2004**, *4*, 986–989.
39. De Maeztu, M.A.; Alava, J.I.; Gay-Escoda, C. Ion implantation: Surface treatment for improving the bone integration of titanium and Ti6Al4V dental implants. *Clin. Oral Implant. Res.* **2003**, *14*, 57–62.
40. Kusy, R.P.; Tobin, E.J.; Whitley, J.Q.; Sioshansi, P. Frictional coefficients of ion-implanted alumina against ion-implanted beta-titanium in the low load, low velocity, single pass regime. *Dent. Mater.* **1992**, *8*, 167–172.
41. Rapisarda, E.; Bonaccorso, A.; Tripi, T.R.; Condorelli, G.G.; Torrisi, L. Wear of nickel-titanium endodontic instruments evaluated by scanning electron microscopy: Effect of ion implantation. *J. Endod.* **2001**, *27*, 588–592.
42. Sawase, T.; Yoshida, K.; Taira, Y.; Kamada, K.; Atsuta, M.; Baba, K. Abrasion resistance of titanium nitride coatings formed on titanium by ion-beam-assisted deposition. *J. Oral Rehabil.* **2005**, *32*, 151–157.
43. Tamura, Y.; Yokoyama, A.; Watari, F.; Kawasaki, T. Surface properties and biocompatibility of nitrided titanium for abrasion resistant implant materials. *Dent. Mater. J.* **2002**, *21*, 355–372.
44. Gil, F.J.; Solano, E.; Campos, A.; Boccio, F.; Sáez, I.; Alfonso, M.V.; Planell, J.A. Improvement of the friction behaviour of NiTi orthodontic archwires by nitrogen diffusion. *Biomed. Mater. Eng.* **1998**, *8*, 335–342.
45. Gil, F.J.; Solano, E.; Mendoza, A.; Pena, J. Inhibition of Ni release from NiTi and NiTiCu orthodontic archwires by nitrogen diffusion treatment. *J. Appl. Biomater. Biomech.* **2004**, *2*, 151–155.
46. Tripi, T.R.; Bonaccorso, A.; Condorelli, G.G. Fabrication of hard coatings on NiTi instruments. *J. Endod.* **2003**, *29*, 132–134.

47. Alsabeeha, N.H.; Swain, M.V.; Payne, A.G. Clinical performance and material properties of single-implant overdenture attachment systems. *Int. J. Prosthodont.* **2011**, *24*, 247–254.
48. Yi, K.J.; Kim, S.G.; Moon, S.Y.; Lim, S.C.; Son, J.S.; Kim, C.G.; Chung, K. Vertical distraction osteogenesis using a titanium nitride-coated distractor. *Oral Surg. Oral Med. Oral Pathol. Oral Radiol. Endod.* **2009**, *107*, e5–e9.
49. Abi Nader, S.; de Souza, R.F.; Fortin, D.; de Koninck, L.; Fromentin, O.; Albuquerque, R.F., Jr. Effect of simulated masticatory loading on the retention of stud attachments for implant overdentures. *J. Oral Rehabil.* **2011**, *38*, 157–164.
50. Assunção, W.G.; Delben, J.A.; Tabata, L.F.; Barão, V.A.; Gomes, E.A.; Garcia, I.R., Jr. Preload evaluation of different screws in external hexagon joint. *Implant Dent.* **2012**, *21*, 46–50.
51. Wichelhaus, A.; Geserick, M.; Hibst, R.; Sander, F.G. The effect of surface treatment and clinical use on friction in NiTi orthodontic wires. *Dent. Mater.* **2005**, *21*, 938–945.
52. D'Antò, V.; Rongo, R.; Ametrano, G.; Spagnuolo, G.; Manzo, P.; Martina, R.; Paduano, S.; Valletta, R. Evaluation of surface roughness of orthodontic wires by means of atomic force microscopy. *Angle Orthod.* **2012**.
53. Husmann, P.; Bourauel, C.; Wessinger, M.; Jäger, A. The frictional behavior of coated guiding archwires. *J. Orofac. Orthop.* **2002**, *63*, 199–211.
54. Kusy, R.P.; Whitley, J.Q.; de Araújo Gurgel, J. Comparisons of surface roughnesses and sliding resistances of 6 titanium-based or TMA-type archwires. *Am. J. Orthod. Dentofac.* **2004**, *126*, 589–603.
55. Cash, A.; Curtis, R.; Garrigia-Majo, D.; McDonald, F. A comparative study of the static and kinetic frictional resistance of titanium molybdenum alloy archwires in stainless steel brackets. *Eur. J. Orthodont.* **2004**, *26*, 105–111.
56. Neumann, P.; Bourauel, C.; Jäger, A. Corrosion and permanent fracture resistance of coated and conventional orthodontic wires. *J. Mater. Sci. Mater. Med.* **2002**, *13*, 141–147.
57. Peitsch, T.; Klocke, A.; Kahl-Nieke, B.; Prymak, O.; Epple, M. The release of nickel from orthodontic NiTi wires is increased by dynamic mechanical loading but not constrained by surface nitridation. *J. Biomed. Mater. Res. A* **2007**, *82*, 731–739.
58. Kim, H.; Johnson, J.W. Corrosion of stainless steel, nickel-titanium, coated nickel-titanium, and titanium orthodontic wires. *Angle Orthod.* **1999**, *69*, 39–44.
59. Schuster, G.; Reichle, R.; Bauer, R.R.; Schopf, P.M. Allergies induced by orthodontic alloys: Incidence and impact on treatment-Results of a survey

in private orthodontic offices in the Federal State of Hesse, Germany. *J. Orofac. Orthop.***2004**, *65*, 48–59.

60. Pernier, C.; Grosgogeat, B.; Ponsonnet, L.; Benay, G.; Lissac, M. Influence of autoclave sterilization on the surface parameters and mechanical properties of six orthodontic wires. *Eur. J. Orthodont.* **2005**, *27*, 72–81.
61. Ryan, R.; Walker, G.; Freeman, K.; Cisneros, G.J. The effects of ion implantation on rate of tooth movement: An *in vitro*model. *Am. J. Orthod. Dentofac.* **1997**, *112*, 64–68.
62. Cobb, N.W.; Kula, K.S.; Phillips, C.; Proffit, W.R. Efficiency of multi-strand steel, superelastic Ni-Ti and ion-implanted Ni-Ti archwires for initial alignment. *Clin. Orthod. Res.* **1998**, *1*, 12–19.
63. Kula, K.; Phillips, C.; Gibilaro, A.; Proffit, W.R. Effect of ion implantation of TMA archwires on the rate of orthodontic sliding space closure. *Am. J. Orthod. Dentofac.* **1998**, *114*, 577–580.
64. Sisera, M.; Hofer, D.J.; Sener, B.; Attin, T.; Schmidlin, P.R. *In vitro* evaluation of three curettes with edge retention technology after extended use. *Schweiz Monatsschr Zahnmed.* **2009**, *119*, 1200–1208.
65. Aspriello, S.D.; Piemontese, M.; Levrini, L.; Sauro, S. Ultramorphology of the root surface subsequent to hand-ultrasonic simultaneous instrumentation during non-surgical periodontal treatments: An *in vitro* study. *J. Appl. Oral Sci.* **2011**, *19*, 74–81.
66. Steele, J.G.; McCabe, J.F.; Barnes, I.E. Properties of a titanium nitride coating for dental instruments. *J. Dent.* **1991**, *19*, 226–229.
67. Rawlings, R.D.; Robinson, P.B.; Rogers, P.S. The durability of ceramic coated dental instruments. *Eur. J. Prosthodont. Restor. Dent.* **1995**, *3*, 211–216.
68. Jackson, M.J.; Sein, H.; Ahmed, W.; Woodwards, R. Novel diamond-coated tools for dental drilling applications. *J. Med. Eng. Technol.* **2007**, *31*, 81–93.
69. Andreasen, G.F.; Hilleman, T.B. An evaluation of 55 cobalt substituted Nitinol wire for use in orthodontics. *J. Am. Dent. Assoc.* **1971**, *82*, 1373–1375.
70. Goldberg, J.; Burstone, C.J. An evaluation of beta titanium alloys for use in orthodontic appliances. *J. Dent. Res.* **1979**,*58*, 593–599.
71. Walia, H.; Brantley, W.A.; Gerstein, H. An initial investigation of the bending and torsional properties of Nitinol root canal files. *J. Endod.* **1988**, *14*, 346–351.

Chapter 9

Microstructural Characteristics and Tribological Behavior of HVOF-Sprayed Novel Fe-Based Alloy Coatings

Andrea Milanti[1,*], *Heli Koivuluoto*[1], *Petri Vuoristo*[1], *Giovanni Bolelli*[2], *Francesco Bozza*[2] *and Luca Lusvarghi*[2]

[1] *Department of Materials Science, Tampere University of Technology, Tampere 33720, Finland*
[2] *Department of Engineering "Enzo Ferrari", University of Modena and Reggio Emilia, Modena 41122, Italy*

ABSTRACT

Thermally-sprayed Fe-based coatings have shown their potential for use in wear applications due to their good tribological properties. In addition, these kinds of coatings have other advantages, e.g., cost efficiency and positive environmental aspects. In this study, the microstructural details and tribological performances of Fe-based coatings (Fe-Cr-Ni-B-C and Fe-Cr-Ni-B-Mo-C) manufactured by High Velocity Oxygen Fuel (HVOF) thermal spray process are evaluated. Traditional Ni-based (Ni-Cr-Fe-Si-B-C) and hard-metal (WC-CoCr) coatings were chosen as references. Microstructural investigation (field-emission scanning electron microscope FESEM and X-Ray diffractometry XRD) reveals a high density and low oxide content for HVOF Fe-based coatings. Particle melting and rapid solidification resulted in a metastable austenitic phase with precipitates of mixed carbides and borides of chromium and iron which lead to remarkably high nanohardness. Tribological performances were evaluated by means of the ball on-disk dry sliding wear test, the rubber-wheel dry particle abrasion test, and the cavitation erosion wear test. A higher wear resistance validates Fe-based coatings as a future alternative to the more expensive and less environmentally friendly Ni-based alloys.

Keywords: thermal spray; HVOF; iron-based coating; microstructure; sliding wear; abrasion wear; cavitation erosion wear

1. INTRODUCTION

Wear is among the most serious of issues to deal with when contacting surfaces in machine parts are in relative motion, such as in bearings, joints, gears, shafts, hydraulic parts, mining industry machines, components in aeronautical industry and others. A 1997 UK industry survey estimated the cost of wear in UK industry to be about £650 million besides overhaul and machine turnover [1]. Therefore, great effort has been undertaken to minimise wear cost by design or material changes. In such a scenario, surface coating technology has gained growing interest in recent years. Among various surface coating techniques, thermal spraying has been recognized as one of the most feasible and cost-effective solutions to protect against wear and corrosion [2]. The High Velocity Oxygen Fuel (HVOF) spray process developed in the 1970s is well known as the standard technology for depositing high wear and corrosion resistant coatings. Fuel gas (propane, ethene, hydrogen) or liquid (kerosene) are mixed in a combustion chamber with oxygen at high pressure. The exhaust hot gases generated by the combustion are forced to pass through a convergent-divergent nozzle reaching supersonic velocity. Powders with typical particle sizes of 5–45 μm are axially injected into the stream of gases and dragged at supersonic velocity towards a substrate [3]. The suitable combination of moderate process temperature and high particle velocity made HVOF the standard process for manufacturing highly dense cermets coatings (e.g., WC-based and Cr_3C_2-NiCr) with excellent adhesion to the substrate and low thermal deterioration [4]. Low porosity, high hardness, high adhesion strength and compression stress are among the most important properties of HVOF-sprayed Ni-based or WC-based coatings which, in turn, comply with the requirement for high wear resistance. Nevertheless, the high and fluctuating prices of precious strategic metals such as Ni and of compounds such as WC are also driving forces for research and development to find solutions to replace such traditional coatings. Additionally, several diseases/pathologies related to nickel (pulmonary and respiratory disease—asthma, pneumoconiosis, cancer; skin diseases—eczema, allergen [5]) and WC-based materials (potential lung disease and carcinogenic agents [6]) have been recently pointed out.

Thermally sprayed Fe-based coatings have been extensively studied as a valid alternative to conventional Ni- and WC-based coatings when high anti-wear and corrosion properties are needed. The satisfactory combination of high mechanical and chemical properties, the low cost, the low environmental impact together with the health and safety make these coatings a reliable future candidate in replacing conventional thermal sprayed coatings. Excellent anti-wear and corrosion properties of Fe-based coatings were achieved by several authors [7,8,9,10]. Moreover the development of alloy compositions with high Glass Forming Ability (GFA) combined with high kinetic spraying process such as HVOF allowed for deposition of amorphous Fe-based coatings with outstanding wear resistance and corrosion properties [11,12]. Although promising results in wear and corrosion applications by using thermally sprayed Fe-based coatings have been achieved, no systematic research on relation between microstructure, mechanical properties and wear behavior of Fe-based coatings has been performed yet.

The present work focuses on the characteristics and wear properties of HVOF-sprayed Fe-Ni-Cr-B-C and Fe-Ni-Cr-Mo-B-C coatings. The chemical composition of the present Fe-based powders includes large amounts of Cr in order to form hard fine-grained precipitates, Ni to form an austenitic matrix with superior mechanical and corrosion properties compared to ferritic structure, and B as hardening element by promoting the formation of an amorphous phase and by developing hard fine-grained precipitates [13,14]. The second composition was achieved by a customized addition of Mo in order to evaluate its effect on the properties of the resulting coatings. Several authors have reported the beneficial effect of lubricious oxides such as MoO_3 and $NiMoO_4$ on reducing the friction coefficient and, consequently, on improving sliding wear behavior [15,16,17]. Two conventional HVOF-sprayed Ni-based and WC-based coatings were chosen as reference. The aim of this work is to study HVOF-sprayed Fe-based coatings as a valid alternative to conventional and expensive thermally-sprayed cermet and Ni- and Co-based coatings as a wear resistant coating solution.

2. EXPERIMENTAL PROCEDURE

2.1. Materials and Coating Manufacturing

Two commercially available gas atomized Fe-based powders, Fe-Ni-Cr-B-C with particle size of −45 + 15 μm and Fe-Ni-Cr-Mo-B-C with particle size of −40 + 20 μm (Wall Colmonoy Ltd., Pontardawe, UK), were employed as feedstock powders. Additionally, three powders were chosen as reference materials as representative state-of-art anti-wear coatings: agglomerated and sintered WC-CoCr (Diamalloy 5834, Sulzer Metco, Wohlen, Switzerland) with particle size of −38 + 15 μm, sintered and crushed WC-CoCr (Amdry 5843, Sulzer Metco, Wohlen, Switzerland) with particle size of −45 + 16 μm and gas atomized Ni-Cr-Fe-Si-B-C (Diamalloy 2001, Sulzer Metco, Wohlen,

Switzerland) with particle size of −45 + 15 μm. The nominal chemical compositions of the powders are listed in Table 1.

Table 1. Nominal chemical composition of powders (wt%) given by supplier.

Element	Powder				
	Fe-Ni-Cr-B-C	Fe-Ni-Cr-Mo-B-C	Ni-Cr-Fe-Si-B-C	WC-CoCr_1 (Diamalloy 5834)	WC-CoCr_2 (Amdry 5843)
Fe	bal.	bal.	4	–	–
Ni	9	12	bal.	1	–
Cr	32	31	17	4	4
Mo	–	2	–	–	–
B	4	3.6	3.5	–	–
C	0.6	0.6	1	–	–
Si	–	–	4	–	–
WC	–	–	–	85	86
Co	–	–	–	10	10

Sulzer Metco Diamond-Jet Hybrid 2700 torch (Sulzer Metco, Wohlen, Switzerland) was selected to spray the powders onto low carbon steel substrates at Tampere University of Technology. Substrates were grit-blasted by using alumina grit (36 mesh) prior to spraying (grit-blasted substrate roughness R_a = 8.3 ± 0.9 μm). The process parameters for the deposition of the Fe-based and of the reference coatings (WC-CoCr, Ni-Cr-Fe-Si-B-C) are presented in Table 2.

Table 2. High velocity oxygen fuel (HVOF) process parameters for the DJ Hybrid 2700 torch.

Parameters	Powder				
	Fe-Ni-Cr-B-C	Fe-Ni-Cr-Mo-B-C	Ni-Cr-Fe-Si-B-C	WC-CoCr_1	WC-CoCr_2
Air flow rate (SL/min)	375	375	375	375	375
O_2 flow rate (SL/min)	190	240	190	240	240
Propane flow rate (SL/min)	70	70	70	70	70
Powder feed rate (g/min)	60	60	60	60	60
Spray distance (mm)	230	230	230	230	230

2.2. Microstructural and Mechanical Characterization

The microstructure of the powders was investigated by Scanning Electron Microscope (SEM, Philips XL30, FEI, Eindhoven, The Netherlands) equipped with Energy Dispersive X-ray (EDX) microanalysis and their phase composition was assessed by X-Ray Diffractometry (XRD, Empyrean, PANAlytical, Cu-Kα radiation, The Netherlands). Experimental conditions include 2θ range 20–100°, step size 0.02°, scan speed 0.02°/s, fixed incident beam mask 10 mm, irradiated length 5 mm, Nickel filter and PANAlytical PIXcel 3D detector. Phase identification was performed using the PANAlytical X'Pert High Score Plus software using the ICDD JCPDF-2 database (International Centre for Diffraction Data, Newtown Square, PA, USA). Particle size distribution and thermal oxidation behavior were studied by a laser diffraction method using a wet dispersion technique (Sympatec laser diffraction system) and by thermogravimetry analysis in air (TG: DTA 404, Netzsch, Selb, Germany), respectively. The microstructure of HVOF-sprayed coatings was observed by using Field-Emission Scanning Electron Microscope (FESEM, Carl Zeiss Microscopy GmbH, Germany) equipped with EDX microanalysis (INCA, Oxford instruments, High Wycombe, UK) and by XRD, under the same experimental conditions mentioned above.

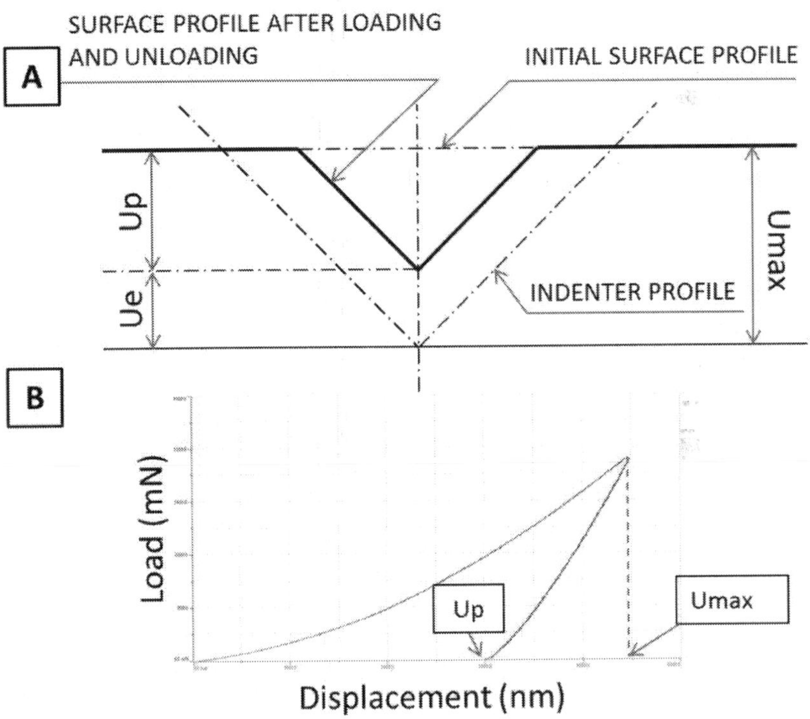

Figure 1. Schematic representation of (**A**) indented surface and (**B**) actual load-displacement curve.

Depth-sensing Berkovich micro- and nano-indentation tests (Micro-Combi tester, CSM instruments, Peseux, Switzerland) were also performed on the polished cross sections (mirror-like surface finish $R_a < 0.05$ μm) in order to evaluate the interlamellar and intralamellar hardness (intrinsic hardness of sprayed particle) respectively as average of 20 measurements [18]. Nano-indentations were purposely placed avoiding any particle boundaries in order to assess intralamellar hardness. Micro-indentations were performed with maximum load of 3 N, loading/unloading rate of 4.5 N/min and holding time at maximum load of 15 s. Nano-indentations were performed with maximum penetration depth of 300 nm, loading/unloading rate of 25 mN/min and holding time at maximum depth of 10 s. Nano- and Micro- Vickers hardness was calculated using the Oliver-Pharr procedure [19] in accordance with ISO 14577 standards [20]. Coating surface elasticity was estimated by the parameter $\eta_{IT} = W_{el}/W_{tot}$ defined as the ratio of the elastic work of indentation (computed as the area under the unloading part of the load-penetration curve) and the total work of indentation (area under the loading part of the curve) as shown in Figure 1.

Bond strength values were determined according to the standard EN582 [21] in a tensile pull test (Instron 1185 mechanical testing machine, Norwood, MA, USA). The glue used was FM-1000 glue foil. Three measurements were carried out to calculate the average values of bond strengths. Load was measured and bond strength was calculated as load per area (diameter of the sample was 25 mm).

2.3. Wear Test

2.3.1. Dry Sliding Wear Resistance

Dry sliding wear resistance of coatings was assessed by a pin-on-disk tribometer with a ball-on-disk configuration. The sample was embedded in phenolic resin, polished up to a mirror-like surface finish ($R_a < 0.05$ μm) and fixed in the middle of a still table. A spherical alumina monocrystalline ball of 6.35 mm in diameter (nominal hardness ≈ 22 GPa), mounted on a rotating pin holding arm, is pressed onto the sample surface with 5 N normal load. Tests were carried out at room temperature, the wear track radius was 3.8 mm, arm the revolution speed was 150 rpm and the test lasted for 50,000 laps. The friction coefficient was continuously measured throughout the test. The wear track was measured through an optical profilometer (WYKO NT1100, Veeco, Tucson, AZ, USA) in order to calculate volume loss. The wear scars were also inspected by SEM.

Dry sliding wear rate is calculated following Equation (1):

$$K\ [\mathrm{mm^3/Nm}] = \frac{V}{w \cdot s} \qquad (1)$$

where V is the volume loss (mm^3), w is the normal load (N) and s is the sliding distance (m).

Alumina counterpart wear rate was measured as well.

2.3.2. Abrasion Wear Resistance

Abrasion wear resistance was assessed by a modified version of the ASTM G65 [22] dry rubber-wheel abrasion test. Sample surfaces were ground by using 1200 grit SiC paper ($R_a \approx 8$ μm) and afterwards pressed with a normal load of 23 N against a rotating rubber wheel with a surface speed of 1.64 m/s. The test lasted for 60 min for an overall wear distance of 5904 m. Every 12 min period, the sample weight loss was measured using scales with 0.001 g accuracy and the sample position was changed to limit possible differences in wear characteristic between different positions. Dry quartz sand (SiO_2) with a grain size ranging from 0.1 to 0.6 mm was used as the abrasive. The flow rate of the abrasive was 25 g/min. The coating wear rate after a one hour test was calculated in accordance with Equation (1).

2.3.3. Cavitation Erosion Wear Resistance

Cavitation erosion wear resistance was evaluated using a vibratory apparatus according to ASTM G32 [23]. The tip of a vibrating horn, connected to a piezoelectric ultrasonic transducer, pulses in a distilled water bath at a frequency of 20 kHz with an amplitude of 50 μm. A high concentration of micrometric bubbles is produced in the fluid kept at a temperature of about 25 °C. The bubbles impact against the sample surface, polished up to a mirror-like surface finish ($R_a < 0.05$ μm) and placed at 1 mm distance from the vibrating tip. The weight loss is periodically measured after 15 min during the first hour (in order to evaluate the incubation periods: at the very initial stage of test, material may be work-hardened and micro-cracks may form without any loss of material) and afterwards every hour up to 5 h of test. The weight loss is converted to a mean depth of erosion (MDE) following Equation (2):

$$MDE\ [\mu m] = \frac{10\Delta W}{\rho A} \tag{2}$$

where ΔW is the weight loss (mg), A is the worn area (cm^2) and ρ is the density of the material tested. Once the cavitation rate has reached a steady value, the cavitation resistance (R_e) is calculated as:

$$R_e\ [h/\mu m] = (SER)^{-1} \tag{3}$$

where SER stands for the steady erosion rate calculated as the slope of the linear fitting of MDE(t) when it becomes steady. The cavitation erosion resistance of a bulk Monel 400 alloy, generally used in maritime and hydraulic components (ρ = 8.8 g/cm^3), was also assessed as a reference.

3. RESULTS AND DISCUSSION

3.1. Microstructural Characterization of Fe-Based Powders

A spherical shape with a smooth surface and little amount of satellites was observed for both Fe-based powders (Figure 2). This is indicative of favorable fluidity during powder feeding. Fe-Ni-Cr-Mo-B-C had narrower particle size distribution (comprised within the −40 + 20 μm range) compared to Fe-Ni-Cr-B-C (comprised within the −50 + 20 μm range) (Figure 3). Both powders showed a similar thermal behavior (Figure 4). TG analysis performed in air revealed a noticeable gain of weight starting at about 1000 °C due to oxidation.

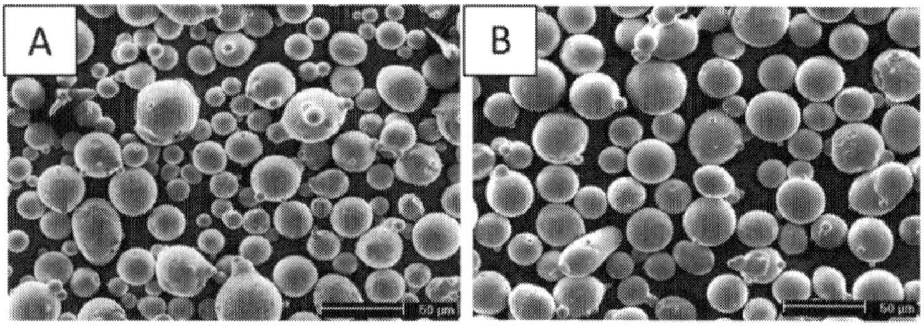

Figure 2. Morphology of (**A**) Fe-Ni-Cr-B-C and (**B**) Fe-Ni-Cr-Mo-B-C powders. Secondary electron SEM images.

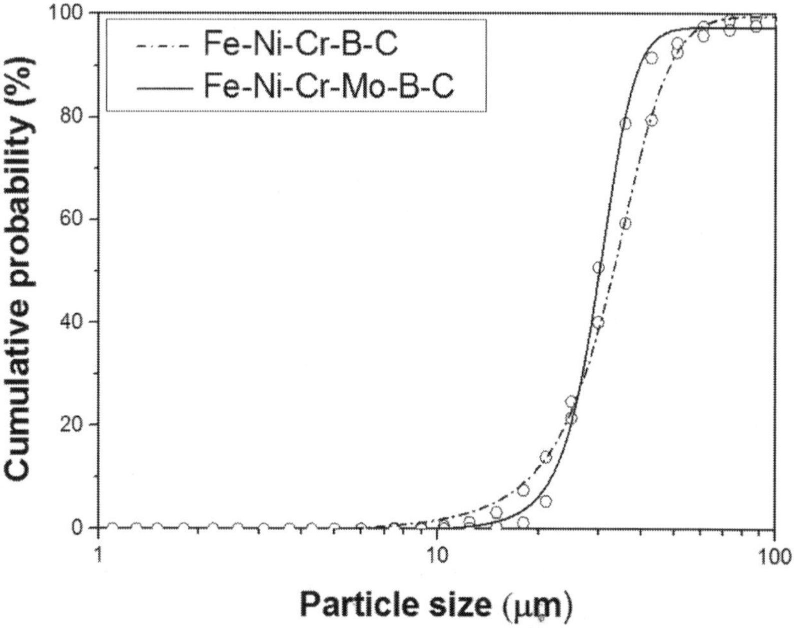

Figure 3. Particle size distributions of Fe-Ni-Cr-B-C and Fe-Ni-Cr-Mo-B-C powders. Measured by using laser diffraction.

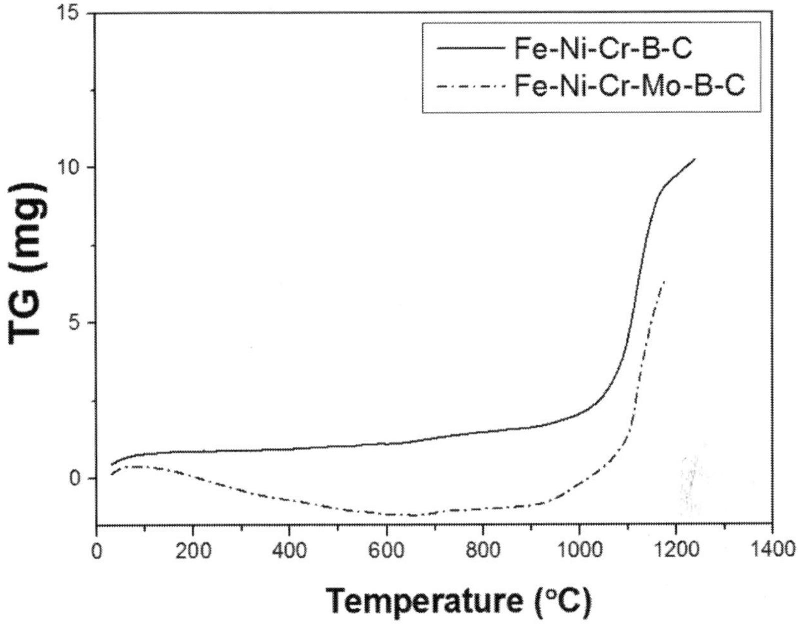

Figure 4. Thermogravimetry (TG) analysis of Fe-Ni-Cr-B-C and Fe-Ni-Cr-Mo-B-C powders.

Cross-sectional SEM micrographs (Figure 5A,B) and EDS analysis (Figure 5C,D) of both powders showed heterogeneous microstructure consisting of acicular, long-shaped Cr-rich precipitates embedded and evenly distributed in a Fe-rich matrix. Cr-enriched compounds (compared to the surrounding area) are clearly shown in EDX analysis (spectra 1, 2 Figure 5C,D). XRD patterns (Figure 6) evaluation revealed the presence of austenitic (f.c.c.) ϒ matrix (JCPDS 00-03-0397) and several low intensity peaks were assigned to chromium and iron mixed carbides and borides (Cr_3C_2 03-065-2427, Cr_7C_3 01-089-5902, $(Cr,Fe)_{23}C_6$ JCPDS 01-078-1500, $(Cr,Fe)_2B$ JCPD 01-072-1073). $(Cr,Fe)_2B$, in particular, seems to be the most abundant of the secondary phases, while the other ones appear in very low amounts. Most of the acicular Cr-rich precipitates probably correspond to this phase.

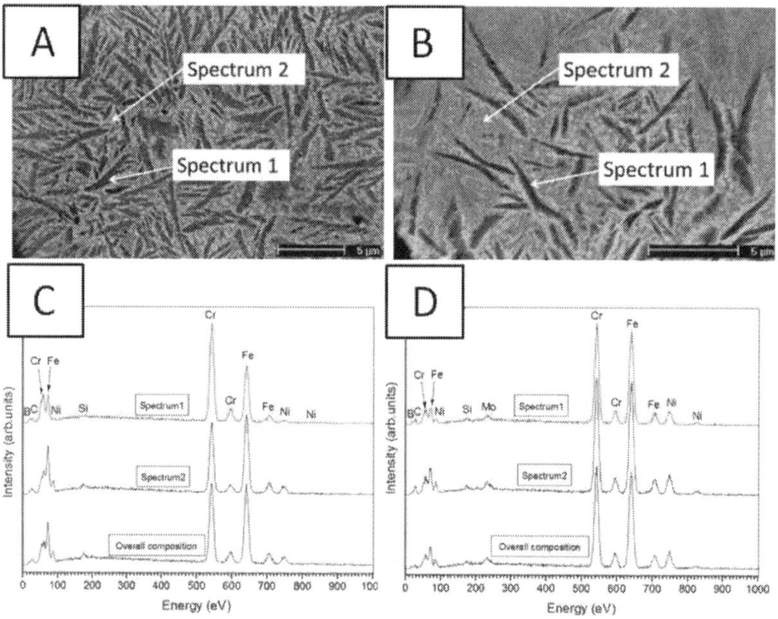

Figure 5. Cross-sectional structure (Backscattered electron SEM images) and Energy Dispersive X-ray (EDX) analysis of (**A**,**C**) Fe-Ni-Cr-B-C and (**B**,**D**) Fe-Ni-Cr-Mo-B-C coatings.

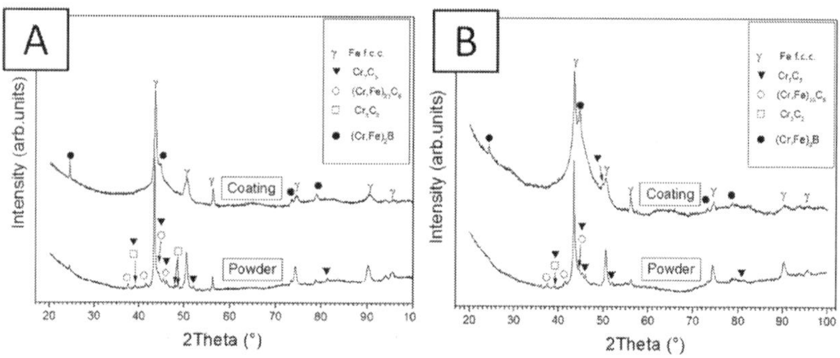

Figure 6. X-Ray diffractometry (XRD) patterns of (**A**) Fe-Ni-Cr-B-C and (**B**) Fe-Ni-Cr-Mo-B-C powders and of the respective coatings.

3.2. Microstructural and Mechanical Characterization of Coatings

Fe-based coatings showed quite dense microstructure (coating cross-sections in Figure 7A,B) with a calculated porosity lower than 1 vol% (image analysis) and a thickness of about 250 μm. Both microstructures were characterized by a high particle flattening ratio with some rounded particles. Remarkable microstructural differences between Fe-Ni-Cr-B-C and Fe-NiCr-Mo-B-C coatings are shown by FESEM characterization (coating cross-sections in Figure

7C,D). Fe-Ni-Cr-B-C seems to retain the microstructural details of the corresponding powder (long-shaped acicular precipitates—label "1" in Figure 7C). However, some particles (label "2" in Figure 7C) present fine-grained precipitates likely due to melting of the powder and rapid solidification in a metastable state. Conversely, the microstructure of the Fe-Ni-Cr-Mo-B-C coating showed a higher amount of particles characterized by fine-grained precipitates (label "1" in Figure 7D) and a small development of a precipitate-free particle rim which is thought to consist of a supersaturated solution and/or of an amorphous phase (label "2" in Figure 7D). Only few particles retain precipitates similar to those found in the powder (label "3" in Figure 7D). It is inferred that the Fe-Ni-Cr-Mo-B-C powder particles underwent more extensive melting, due to their narrower particle size distribution (Section 3.1) and to the warmer spraying parameters (Table 2). Therefore, during spraying, carbides and borides are thought to dissolve into the surrounding Fe-rich matrix and subsequently the rapid quenching and solidification results in precipitation of fine-grained Cr- and B-rich compounds embedded into a metastable Fe-rich solid solution which is significantly supersaturated in solute elements. A similar microstructural evolution was previously described in thermally sprayed Fe-Cr-B [24]. Feedstock powder particles consisting of boride-based phases in a Fe-based matrix were indeed also found to turn into featureless lamellae, mainly consisting of a solid solution supersaturated in boron, as a consequence of melting and impact quenching. A higher particle melting of Fe-Ni-Cr-Mo-B-C was also inferred by the lower roughness (R_a= 7.1 ± 0.5 μm) compared to Fe-Ni-Cr-B-C (R_a = 8.0 ± 0.8 μm).

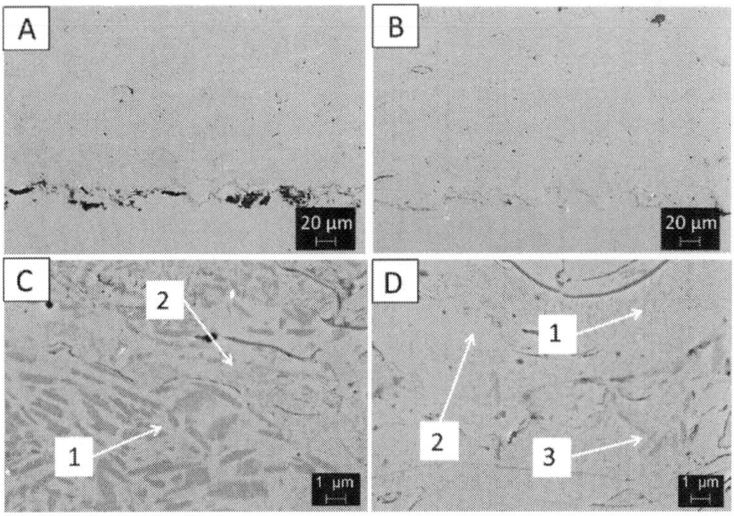

Figure 7. Microstructures of High velocity oxygen fuel (HVOF) (**A**,**C**) Fe-Ni-Cr-B-C and (**B**,**D**) Fe-Ni-Cr-Mo-B-C coatings. Secondary electron FESEM images.

Microstructural changes were confirmed by XRD measurements (Figure 6). The main phase of both coatings was still austenitic (f.c.c.) γ-Fe (JCPDS 00-03-0397) (Figure 6A,B) with some low intensity peaks attributed to chromium and iron mixed boride $(Cr,Fe)_2B$ (JCPD 01-072-1073). Dissolution of the carbides detected in the powders and peak broadening are the results of the extensive microstructural changes occurring from powder to coatings due to the rapid quenching of molten or semi-molten particles at impact. It is worth remembering the effect of B addition. Besides the ability to form fine-grained precipitates, B (4 and 3.6 wt% in Fe-Ni-Cr-B-C and Fe-Ni-Cr-Mo-B-C, respectively) may remarkably increase the viscosity of the molten metal alloy making the melt more stable than the relative crystalline phase [25]. According to the general rules for metallic glass formation, indeed, alloys containing elements with significantly different atomic radii and large negative heats of mixing develop a viscous melt consisting of large atomic clusters, increasing the possibility to form an amorphous phase [26]. Further studies have to be done in order to assert the presence of an amorphous phase (deeper microstructural characterization has to be done, e.g., TEM).

One can clearly observe that the Fe-Ni-Cr-Mo-B-C coating underwent larger microstructural changes compared to Fe-Ni-Cr-B-C. Wider peak broadening at $2\theta \approx 43°$ and the formation of a broad band in the $35° < 2\theta < 50°$ range are observed (Figure 6B). Peak broadening is much likely due to the formation of a finely-structured supersaturated austenitic solid solution, whereas the underlying broad band probably reflects amorphous phase formation [11,12]. In the literature, similar broad bands in the diffraction patterns of thermally sprayed Fe-based alloys have indeed been univocally associated with the presence of significant amounts of amorphous phase, identified e.g., by TEM observation [14]. With regard to Fe-Ni-Cr-Mo-B-C, the narrower particle size distribution and the higher oxygen flow rate, which in turn increases the flame temperature, provide more extensive particle heating and melting, resulting in the observed microstructural changes. Furthermore, Mo is thought to promote high viscosity at high temperature by its stabilizing effect on boride and carbides and in turn it may increase the so called Glass Forming Ability (GFA) of the sprayed alloy [27].

Besides the effect of heating and melting, it is worth mentioning that the peak broadening might be also attributed to high plastic deformation of the particles at impact. Richer *et al.* [28] reported the dissolution of β-NiAl precipitates into Co-Ni-Cr solid solution γ-matrix due to the high energy impact of particles onto the substrate during the deposition of HVOF-sprayed MCrAlY (M = Co,Ni) powders. Most likely a combined effect of melting/rapid quenching and plastic deformation of particles resulted in the dissolution of carbides and borides into the super saturated austenitic matrix. No oxide peaks are recognizable in these patterns.

The micro- and nano- hardness values of the coatings (Figure 8), assessed by depth-sensing Berkovich micro- and nano-indentation tests, are similar to those of a conventional HVOF-sprayed Ni-based alloy, with the exception of the micro-hardness of the Fe-Ni-Cr-B-C sample, which is slightly lower. The high dispersion of the results of WC-CoCr nanohardness was due to the composite microstructure of this material, with fine sized hard carbides embedded in a softer metallic CoCr binder: nanoindentation values therefore reflect the different intrinsic hardnesses of both phases. Instead the high dispersion of the results of Fe-Ni-Cr-B-C nanohardness is due to the heterogeneous microstructure as explained previously. The larger particle size distribution compared to that of Fe-Ni-Cr-Mo-B-C resulted in a less homogeneous microstructure with molten and unmolten particles with different mechanical properties. In fact, the rapid quenching of a molten particle and its solidification in a metastable state may result in greater hardness due to strengthening by solid solution as reported in another study [10]. Since nano-hardness was achieved by indentation at low penetration depth (300 nm) within the particle, it mainly reflects the intralamellar properties of the sprayed coating [29]. Instead the micro-hardness obtained at greater penetration (3 N max load) reflects the interlamellar properties [10,18]. The difference between micro- and nano-hardness is accordingly a measure of interlamellar cohesion, and it grows larger as the cohesion decreases, as shown in the literature [18,30]. It is therefore inferred that, due to the lower degree of melting of the coarser Fe-Ni-Cr-B-C powder, the interlamellar cohesion of this coating is poorer than that of the Fe-Ni-Cr-Mo-B-C coating, where the amount of fully molten particles (recognizable by their higher degree of flattening and by the presence of finer precipitates and/or of precipitate-free metastable solid solutions as described above) is higher.

Bond strengths of the coatings are shown in Table 3. HVOF-sprayed Fe-based coatings possessed bond strengths of about 30 MPa. As the failure of the WC-CoCr_2 after pull test was observed to occur in the glue, its bond strength is higher than the value reported (>64 MPa). The low in-flight oxidation and high specific weight of WC-CoCr which guarantee high kinetic particle velocity play a fundamental role in better particle sticking and in turn adhesion bond strength.

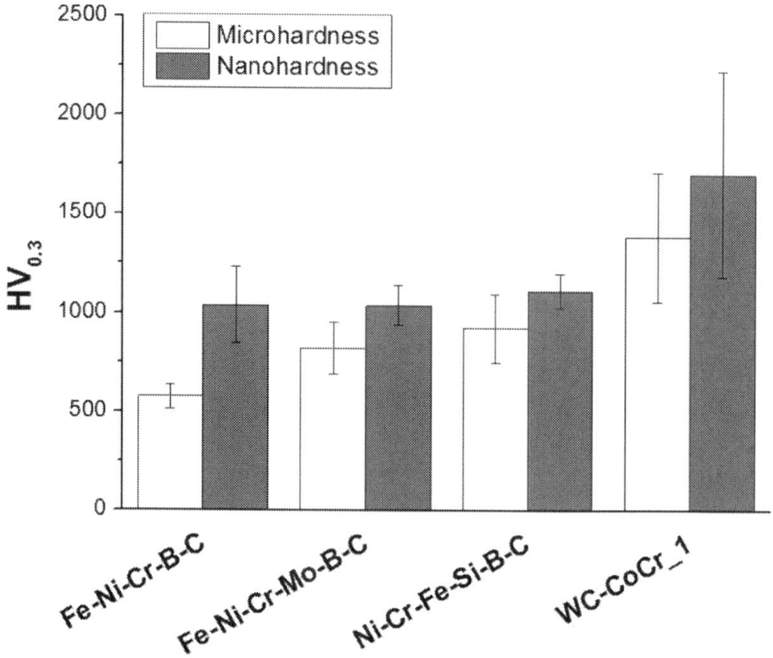

Figure 8. Berkovich micro- and nano- hardness of HVOF-sprayed coatings.

Table 3. Bond strengths and standard deviations of HVOF-sprayed coatings.

Coating	Bond strength (MPa)
Fe-Ni-Cr-B-C	31.9 ± 7.2
Fe-Ni-Cr-Mo-B-C	25.1 ± 4.5

3.3. Wear Properties

3.3.1. Dry Sliding Wear Resistance

Dry sliding wear rates (Table 4) of both Fe-based coatings were found to be lower than those of Ni-Cr-Fe-Si-B-C but still far larger than those of WC-CoCr. Wear rates of Al_2O_3 balls used against Fe-based coatings showed lower values compared to the Al_2O_3/Ni-Cr-Fe-Si-B-C coupling (Table 4). Friction coefficients of Fe-based coatings appeared lower and more stable throughout the test compared to that of Ni-Cr-Fe-Si-B-C (Figure 9). Furthermore, the Fe-Ni-Cr-Mo-B-C coating was slightly more resistant to dry sliding wear showing lower coating wear rates, lower alumina ball wear rates and a lower and more stable friction coefficient compared to Fe-Ni-Cr-B-C.

Table 4. Coating wear rates and Al_2O_3 ball counterpart wear rates after ball-on-disk.

Coating	Sample wear rate ($\times 10^{-6}$ mm^3/Nm)	Alumina ball wear rate ($\times 10^{-6}$ mm^3/Nm)
Fe-Ni-Cr-B-C	3.45 ± 0.81	0.101 ± 0.008
Fe-Ni-Cr-Mo-B-C	3.25 ± 0.44	0.095 ± 0.004
Ni-Cr-Fe-Si-B-C	8.13 ± 0.75	0.203 ± 0.037
WC-CoCr_2	0.06 ± 0.01	1.151 ± 0.131

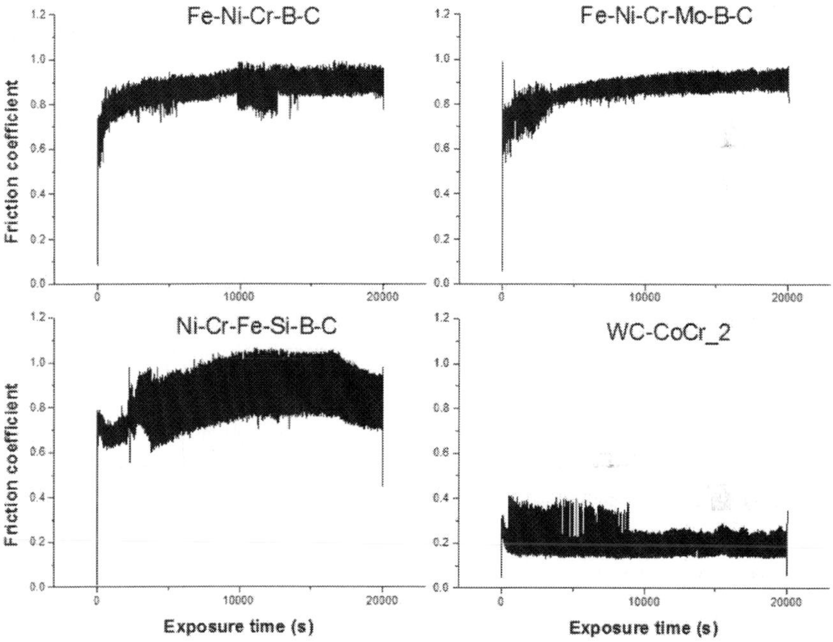

Figure 9. Friction coefficient of HVOF-sprayed coatings against alumina ball recorded throughout the test.

Abrasive wear occurred on both Fe-based coatings (small grooves labelled "1" in Figure 10A,B) due to the relative motion between the softer surface (Fe-based coatings) against a counterpart with harder asperities (alumina). The wear mechanism could have been worsened by the presence of hard agglomerates of oxides trapped between the mating surfaces, formed due to tribo-oxidation phenomena. Adhesive/delamination wear occurred and it was detectable by the presence of a small dark spot likely due to particle pull out (Figure 10A,B: label "2"). During the sliding test micro-welded junctions formed between the coating and the counterpart asperities. The strength of such micro-welded junctions may be greater than the bond strength between particles and this results in particle pull out. Splat delamination may also occur due to the presence of few oxide-rich intersplats which are brittle and cannot accommodate shear strains causing splat pull out [31]. These mechanisms seemed to occur more frequently on the Fe-Ni-Cr-B-C coating, consistent with the previous observations on its lower interlamellar cohesion (Section 3.2). These delaminated areas are covered up with a cluster of fine oxidized debris (spectrum 1 Figure 10C,E), further confirming the existence of triboxidative wear as well. Fe-Ni-Cr-Mo-B-C, by contrast, showed a more evenly distributed oxide layer throughout the wear track (see the dark area in the SEM-BSE inset of Figure 10B, detail in Figure 10D and EDX spectrum in Figure 10F: spectrum 1). The higher amorphous phase content of Fe-Ni-Cr-Mo-B-C coating (previously assumed) may be playing a critical role in oxidation behavior. In fact, as carbides/borides precipitate, large amounts of Cr and Mo are consumed which, in turn, causes depletion of Cr and Mo in the surrounding area and results in a lower capability to form even and continuous oxide scale. The micro-Raman spectrum of this tribo-oxidational layer (Figure 11), characterized by a main peak at 652 cm^{-1}, accordingly confirms a significant presence of Mo. This peak is shifted to a lower wavenumber, compared to pure magnetite [32] and to the main peak in the Raman spectra of tribo-oxidational products of Fe-Ni-Cr-B-C coatings as seen in [10]. Such a shift is reportedly associated with the replacement of Fe with transition metal ions other than Ni and Cr (e.g., Mn, Cu) in the lattice of magnetite [33,34]. Although no specific study on Mo-substituted magnetite is available, the same effect is reasonably expected, leading to the conclusion that this layer contains non-negligible amounts of Mo.

Therefore, such an oxide layer could decrease the shear strength and in turn the frictional forces between the tested surface and the counterpart, reducing adhesive/delamination wear and resulting in a friction coefficient which is not only lower (Table 4), but also more stable (lower fluctuations of the friction curve in Figure 9). It is worth underlining that oxidation may be considered as a desirable event whenever a thin lubricious layer forms [35,36,37]. Further studies are needed to evaluate the positive effect of this oxide layer on coating wear resistance.

The wear track of Ni-Cr-Fe-Si-B-C presented less adhesion/delamination wear, with very limited cracking at the particle boundaries (label “1” in Figure 12A) and few delaminated particles (with fine oxides filling the resulting holes, Figure 12A—spectrum 2). Extensive triboxidation (spectrum 1, Figure 12A) and abrasion are thought to occur. The release of a larger amount of debris fragments in the contact region (due to the higher wear loss) and the instability of the oxide film, testified by the numerous delaminated areas seen in Figure 12A, probably explain the large fluctuations of the friction coefficient produced by sliding wear testing of this sample (Figure 12). Instead very shallow grooves

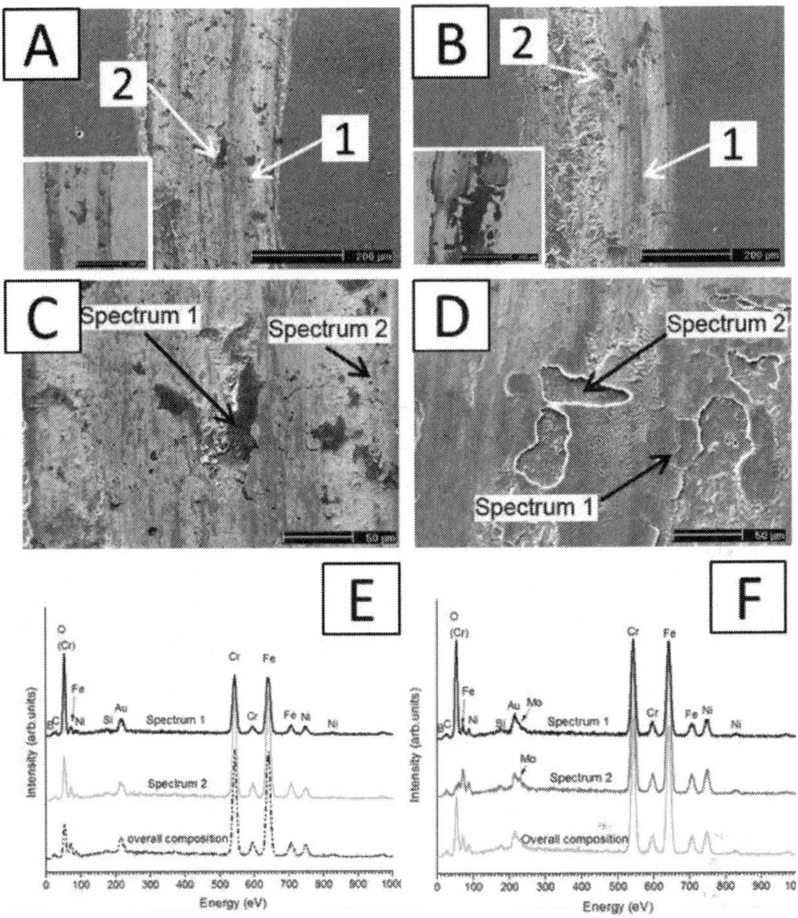

Figure 10. Sliding wear tracks (backscattered electron SEM images) and EDX analysis of (**A**,**C**,**E**) Fe-Ni-Cr-B-C and (**B**,**D**,**F**) Fe-Ni-Cr-Mo-B-C coatings.

(label "1", Figure 12B) and near-zero delaminated areas are detected on the wear track of WC-CoCr. Additionally, consistent with the high wear rate of alumina ball counterpart when testing WC-CoCr, a fine grained alumina particle area is reported (spectrum 1, Figure 12D). The lower friction coefficient of WC-CoCr may further reduce the delamination wear according to Hertzian contact theory. When two non-conformal bodies are in contact with a certain normal load combined with a tangential load (relative movements of contacting surface), the greatest value of maximum shear stress is placed below the contacting surface, at a distance which depends on the relative elastic modulus, on the radius of circular shape, and on the contacting load. As the friction coefficient increases, the maximum shear stress moves towards the surface, and it becomes located on the surface for $\mu > 0.32$ [38]. Furthermore, one can notice the high fluctuation of the WC-CoCr_2 friction coefficient within the first 9000 laps. At the very early stage of the dry sliding wear test, the hard alumina asperities may crack and get dispersed as fine wear debris between two contacting surfaces. Eventually they get stuck into the metal matrix of WC-CoCr_2 (spectrum 1,Figure 12D) resulting in a decrease of the friction coefficient (transfer mechanism, low friction coefficient alumina/alumina is well known and reported in several studies [39,40]).

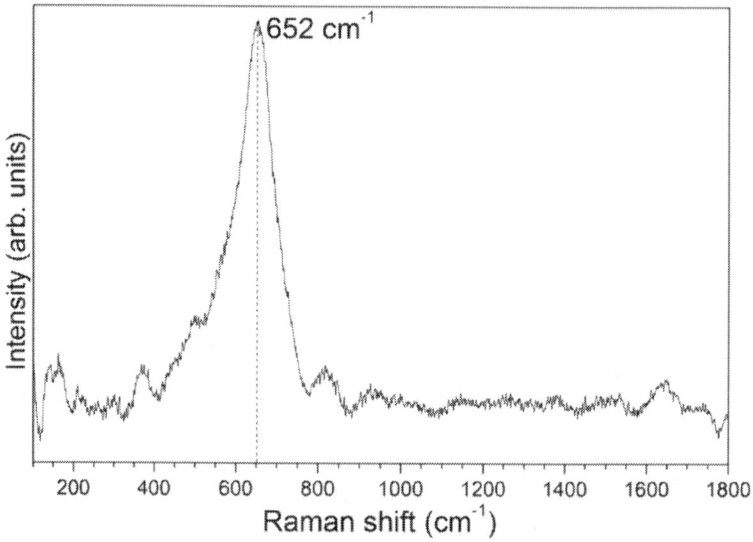

Figure 11. Micro-Raman spectrum of tribo-oxidational layer of Fe-Ni-Cr-Mo-B-C coatings.

3.3.2. Cavitation Erosion Wear Resistance

The cumulative mean depth of erosion (MDE) assessed by cavitation erosion testing in distilled water is shown in Figure 13. WC-CoCr showed the highest wear resistance within the first hour (Figure 13). However, when hard carbides or particles are pulled out of the coatings, severe erosion wear and, in turn, high weight loss is created. Fe-Ni-Cr-Mo-B-C was reported to have the highest cavitation erosion resistance within 5 h (Table 5). In spite of its slightly lower adhesion strength, this sample exhibits better interparticle cohesion than the Fe-Ni-Cr-B-C coating (as witnessed by the lower difference between micro- and nano-hardness values, Section 3.2), which is particularly important for cavitation erosion resistance. In fact, in thermal spray coatings, cavitation wear generally proceeds by propagation of cracks along particle boundaries [8]. While the comparison between micro- and nano-hardness can be significant as a measure of the relative quality of interparticle cohesion, hardness, by itself, is not the most influential property controlling cavitation erosion resistance. This is witnessed by the lower erosion resistance of the hard WC-CoCr coating. Conversely, another mechanism seems to be dominant in this case. In our previous study [41], the remarkable importance of work-hardening and surface elasticity as energy dissipation mechanisms in cavitation erosion wear was discussed. Based on such results, the η_{IT} value (estimation of surface elasticity) was calculated through depth-sensing micro-indentation (Figure 14). A close correlation exists between cavitation erosion resistance and surface elasticity as a key property in order to dissipate the energy released by bubble collapse (Figure 14). The poorer cavitation erosion resistance of Ni-Cr-Fe-Si-B-C and WC-CoCr coatings is indeed well explained by their lower surface elasticity compared to Fe-Ni-Cr-Mo-B-C. The only sample which exhibits lower cavitation erosion resistance in spite of its high elasticity is the Fe-Ni-Cr-B-C coating, on account of its lower interparticle cohesion as observed previously.

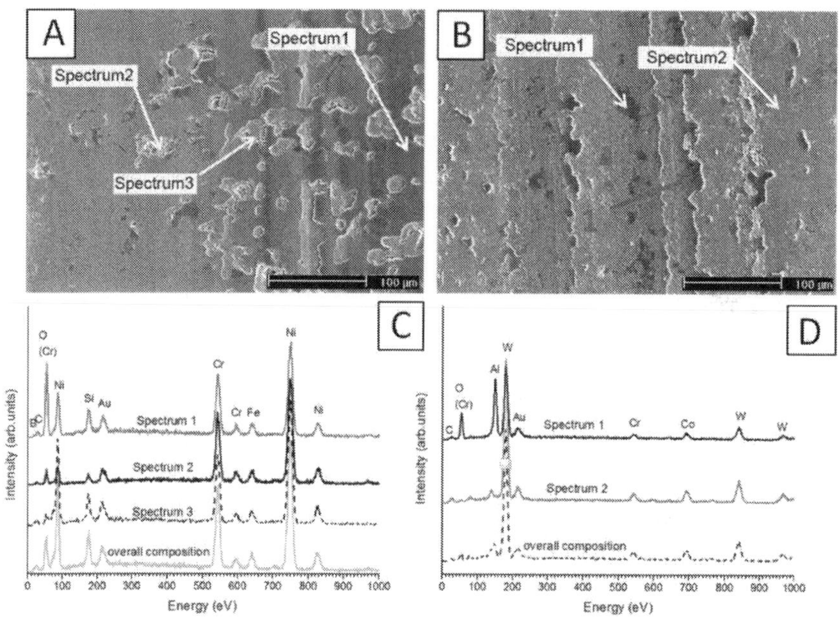

Figure 12. Sliding wear track and EDX analysis of Ni-Cr-Fe-Si-B-C (**A**,**C**) and WC-CoCr_2 (**B**,**D**) coatings.

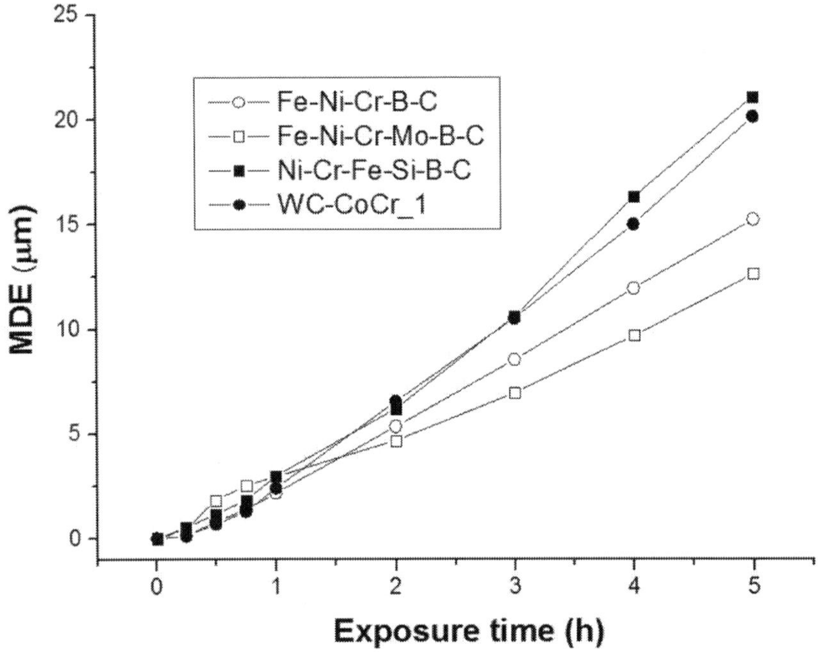

Figure 13. Variation of mean depth of erosion (MDE) of HVOF-sprayed coating samples within 5 h.

Table 5. Steady erosion rate (SER) and cavitation erosion wear resistance (R_e) of HVOF-sprayed coating samples.

Coating	SER (μm/h)	R_e (h/μm)
Fe-Ni-Cr-B-C	3.87	0.26
Fe-Ni-Cr-Mo-B-C	2.43	0.41
Ni-Cr-Fe-Si-B-C	4.62	0.22
WC-CoCr_1	4.39	0.23

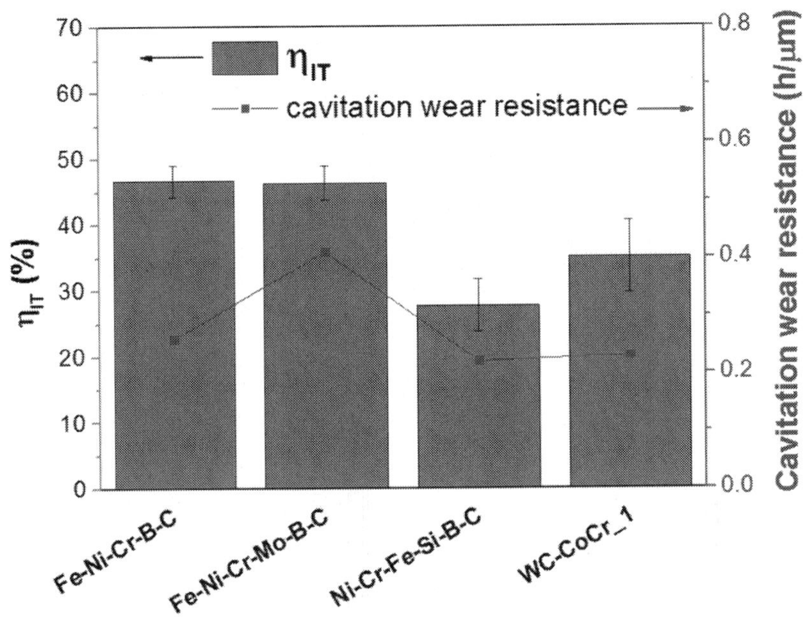

Figure 14. Cavitation erosion wear resistance *vs.* estimation of surface elasticity (η_{IT}) for HVOF-sprayed coating samples.

A further comparison between the best performing Fe-based coating (Fe-Ni-Cr-Mo-B-C) and bulk Monel400 alloy (Ni-30Cu), commonly used for maritime and hydraulic applications (pump shafts, seawater valves, piping system), was performed in order to validate the replacement of conventional bulk material with this promising HVOF Fe-based sprayed coating. In Figure 15, the Ni-30Cu bulk material presented an incubation period within the first hour in which the material is most likely work hardened, or cracks may form without any detectable loss of material. Afterwards, the MDE increased due to crack propagation reaching a steady state. Conversely, Fe-Ni-Cr-Mo-B-C showed higher cavitation wear rate within the first test hour most likely due to the slightly lower strength of the superficial layer of thermal spray coating compared to Monel400 alloy. Indeed, in coating deposited by high kinetic energy thermal spray processes, such as the HVOF technique, extensive deformation and densification occurs due to the "peening effect" of impinging particles that effectively hammer the previously deposited layers [42,43]. The outermost layer is therefore slightly less dense and possesses lower overall strength compared to the inner layers, as it lacks this peening effect. After removal of this superficial layer, however, the cavitation erosion rate of the coating decreased to a lower value. To sum up, the Fe-Ni-Cr-Mo-B-C coating presented a lower steady erosion rate (SER) and in turn higher cavitation wear resistance within the 5 h test.

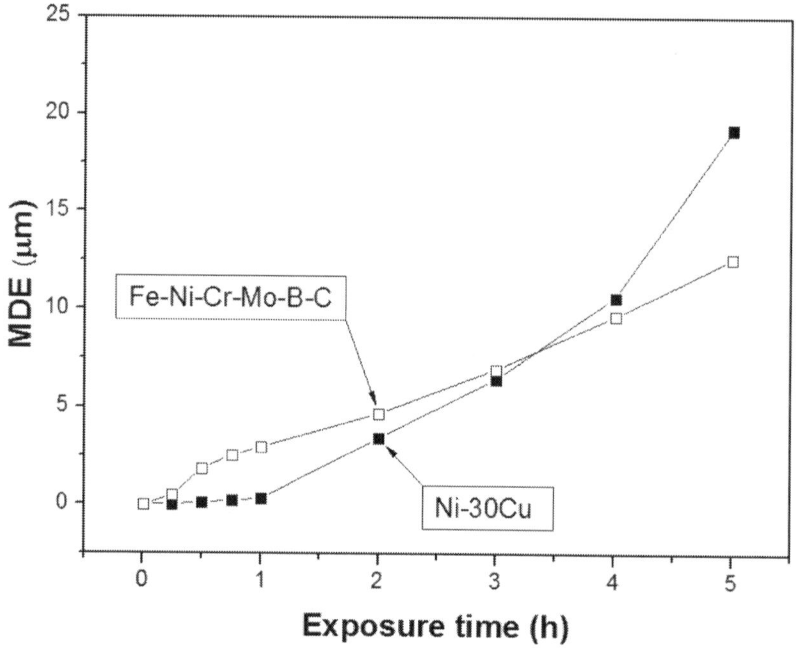

Figure 15. Variation of mean depth of erosion (MDE) of HVOF-sprayed Fe-Ni-Cr-Mo-B-C coating and Ni-30Cu bulk material.

3.3.3. Abrasion Wear Resistance

Abrasion wear resistance was evaluated by the rubber-wheel test. As expected, the WC-CoCr hard metal coating showed a remarkably higher abrasion wear resistance due to its superior hardness. Hardness, which expresses the overall resistance to plastic deformation, plays a very important role in this case. However, Fe-based coatings turned out to be comparable to Ni-Cr-Fe-Si-B-C (Figure 16), or even better, when the weight loss is converted to a volumetric abrasion wear rate accounting for material density (Table 6).

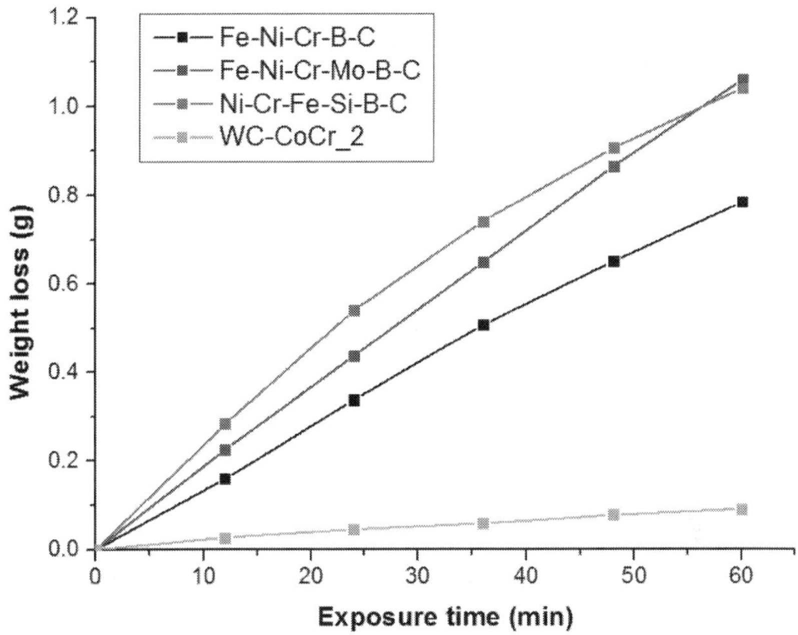

Figure 16. Weight loss of tested sample recorded within a 1-h long rubber-wheel abrasion wear test.

Table 6. Weight loss and abrasion wear rate for a 1-h long rubber-wheel abrasion wear test.

Measured quantity	Coating			
	Fe-Ni-Cr-B-C	Fe-Ni-Cr-Mo-B-C	Ni-Cr-Fe-Si-B-C	WC-CoCr_2
weight loss (mg)	784	1060	1040	90
wear rate (*10^{-5} mm^3/Nm)	74.3	100.5	112.0	4.6

4. CONCLUSIONS

This study focused on the microstructural and mechanical characterization and on the tribological behavior of novel HVOF-sprayed Fe-based coatings compared to conventional HVOF-sprayed Ni-Cr-Fe-Si-B-C and WC-CoCr coatings taken as references. Their microstructural details, micro-mechanical properties, sliding, abrasive and cavitation erosion behavior were studied and to sum up, the main conclusions are listed as follows:

- Fe-based coatings showed a dense microstructure with negligible porosity (<1% calculated through image analysis). The powder microstructure is only partly retained in the coatings after deposition. High flame temperature resulted in in-flight melting of several particles and in their deposition in the molten or semi-molten state. The rapid quenching and solidification led to precipitation of fine-grained Cr- and B-rich compounds embedded into a supersaturated Fe-rich solid solution. Higher microstructural alteration was detected in Fe-Ni-Cr-Mo-B-C due to finer particle size distribution and a warmer set of spraying parameters. No significant oxide inclusions were detected in the coatings.

- Fe-based coatings possess remarkably high nano-hardness revealing high intrinsic hardness of the sprayed material, comparable to that of Ni-Cr-Fe-Si-B-C. Moreover, considering the ratio between micro- and nano- hardness as an estimation of cohesion strength, the Fe-Ni-Cr-Mo-B-C coating was found to have similar micromechanical properties compared to a conventional HVOF-sprayed Ni-based alloy, while the Fe-Ni-Cr-B-C sample had somewhat poorer cohesion. Furthermore, Fe-based coatings possess lower bond strength compared to the reference coatings, as evaluated by tensile testing according to the standard EN582. Bond strength is reported to be about 30 MPa.
- Although the dry sliding wear resistance of HVOF-sprayed Fe-based metal alloy coatings could not compare with that of a cermet material (WC-CoCr), their sliding wear rate was reported to be lower than that of conventional HVOF-sprayed Ni-based metal alloy coatings. Fe-based coatings are found to possess a lower and more stable friction coefficient and more specifically the formation of a Mo-containing oxide scale during sliding wear is effective in reducing adhesion to the counterbody, decreasing friction and lessening the fluctuation of the friction coefficient.
- HVOF-sprayed Fe-based coatings, but most specifically the novel composition (Fe-Ni-Cr-Mo-B-C), presented superior cavitation wear resistance in comparison to those of Ni- and WC-based conventional coatings. Its high surface elasticity, as determined by depth sensing indentation tests, coupled to its good interparticle cohesion, indicated by the relatively low difference between micro- and nano-hardness values, were found to play a fundamental role in the energy dissipation mechanism during bubble impact.
- Abrasion wear resistance of the cermet reference (WC-CoCr) is noticeably higher than that of Fe-based coatings due to higher hardness, which prevents the abrasive sand from penetrating into the coating and from ploughing out a groove by plastic deformation. However, Fe-based coatings showed lower abrasion wear rate compared to a Ni-based reference coating.

As a summary, Figure 17 shows the radar wear rates diagram of HVOF-sprayed Fe-based, Ni-based, WC-CoCr tested coatings. HVOF-sprayed Fe-based coatings represent a valid alternative toNi-based alloys when higher anti-sliding, abrasion and cavitation wear properties are needed due to the high coating quality with acceptable micro-hardness. However, their sliding and abrasion wear resistance could not compare to that of WC-CoCr, whose superior hardness offers higher protection against scratching from hard counterparts, debris and quartz sand. Nevertheless, Fe-based coatings showed outstanding cavitation wear resistance, twice as high as that of WC-CoCr andNi-Cr-Fe-Si-B-C coatings and even better than that of commonly used Ni-30Cu bulk material, mainly due to the higher surface elasticity.

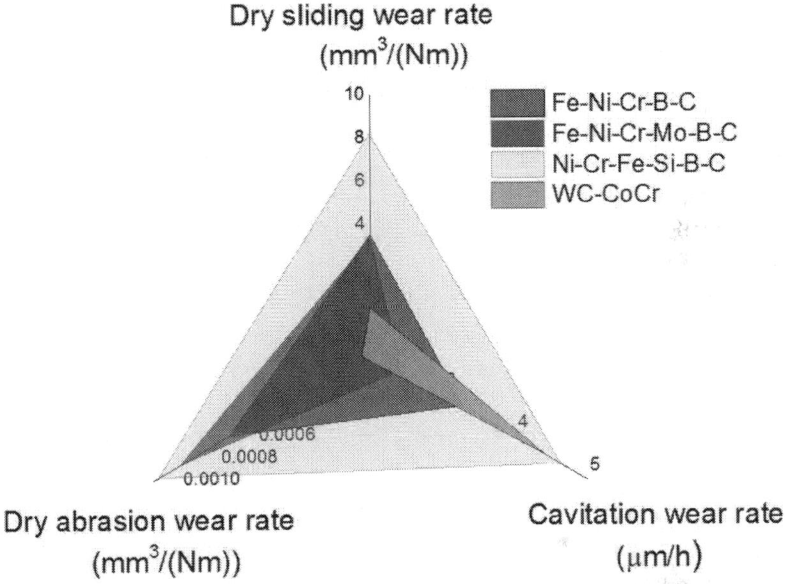

Figure 17. Radar wear rates diagram of HVOF-sprayed Fe-based, Ni-based and WC-CoCr coatings.

The superior surface elasticity of Fe-Cr-Ni-Mo-B-C may be due to the higher amorphous content. However further work involving deeper microstructural analysis has to be done in order to confirm the presence of amorphous phase.

ACKNOWLEDGMENTS

The authors would like to thank Mikko Kylmälahti of Tampere University of Technology, Department of Materials Science for spraying the coatings. The study was supported by the Finnish National Graduate School (Concurrent Mechanical Engineering) and Tampere University of Technology.

REFERENCES

1. Mellor, G.B. *Surface Coatings for Protection against Wear*; CRC Press: Boca Raton, FL, USA, 2006.
2. Davis, J.R. *Handbook of Thermal Spray Technology*; ASM International: Materials Park, OH, USA, 2004; pp. 171–213.
3. Pawlowski, L. *The Science and Engineering of Thermal Spray Coatings*, 2nd ed.; John Wiley & Sons: Chichester, UK, 2008.
4. Berger, L.M. Hardmetal as thermal spray coatings. *Powder Metall.* **2007**, *50*, 205–214.
5. Montavon, G. Safety Issues and Risks Related to Thermal Spray Process, ETSA Summer School Tampere University of Technology, Tampere, Finland, 13–14 June 2012.
6. *Cobalt–Tungsten Carbide: Powders and Hard Metals*; 12th Report on Carcinogens; U.S. Department of Health and Human Services: Washington, DC, USA, 10 June 2011.
7. Yuping, W.; Pinghua, L.; Chenglin, C.; Zehua, W.; Ming, C.; Junhua, H. Cavitation erosion characteristics of a Fe–Cr–Si–B–Mn coating fabricated by high velocity oxy-fuel (HVOF) thermal spray. *Mater. Lett.* **2007**, *61*, 1867–1872.
8. Schwetzke, R.; Kreye, H. *Cavitation Erosion of HVOF Coatings*; ASM International: Material Park, OH, USA, 1996.
9. Pukasiewicz, A.G.M.; Capra, A.R.; Paredes, R.S.C. Cavitation Erosion Mechanism in Fe-Mn-Cr-Si-Ni Arc Thermally Sprayed Coatings. In Proceedings of the International Thermal Spray Conference 2011, Hamburg, Germany, 27–29 September 2011.
10. Bolelli, G.; Bonferroni, B.; Laurila, J.; Lusvarghi, L.; Milanti, A.; Niemi, K.; Vuoristo, P. Micromechanical properties and sliding wear behaviour of HVOF-sprayed Fe-based alloy coatings. *Wear* **2012**, *276–277*, 29–47.
11. Cherigui, M.; Feraoun, H.I.; Feninehe, N.E.; Aourag, H.; Coddet, C. Structure of amorphous iron-based coatings processed by HVOF and APS thermally spraying. *Mater. Chem. Phys.* **2004**, *85*, 113–119.
12. Guo, R.Q.; Zhang, C.; Chen, Q.; Yang, Y.; Li, N.; Liu, L. Study of structure and corrosion resistance of Fe-based amorphous coatings prepared by HVAF and HVOF. *Corros. Sci.* **2011**, *53*, 2351–2356.
13. Kumar, S.; Kim, J.; Kim, H.; Lee, C. Phase dependence of Fe-based bulk metallic glasses on properties of thermal spray coatings. *J. Alloys Compd.* **2009**, *475*, L9–L12.
14. Chokethawai, K.; McCartney, D.G.; Shipway, P.H. Microstructure evolution and thermal stability of an Fe-based amorphous alloy powder and thermally sprayed coatings. *J. Alloys Compd.* **2009**, *480*, 351–359.

15. Shina, J.H.; Wang, Q.M.; Kima, K.H. Microstructural evolution and tribological behavior of Mo–Cu–N coatings as a function of Cu content. *Mater. Chem. Phys.* **2011**, *130*, 870–879.
16. Tiana, B.; Yuea, W.; Fua, Z.; Gu, Y.; Wang, C.; Liu, J. Surface properties of Mo-implanted PVD TiN coatings using MEVVA source. *Appl. Surf. Sci.* **2013**, *280*, 482–488.
17. Woydt, M.; Skopp, A.; Dörfel, I.; Witk, K. Wear engineering oxides/anti-wear oxides. *Wear* **1998**, *218*, 84–95.
18. Nohava, J.; Bonferroni, B.; Bolelli, G.; Lusvarghi, L. Interesting aspects of indentation and scratch methods for characterization of thermally-sprayed coatings. *Surf. Coat. Technol.* **2010**, *205*, 1127–1131.
19. Archard, J.F. Contact and rubbing of flat surface. *J. Appl. Phys.* **1953**, *24*, 981–988.
20. *ISO 14577-1:2002(en): Metallic Materials—Instrumented Indentation Test for Hardness and Materials Parameters*; International Organization for Standardization: Geneva, Switzerland, 2002.
21. *EN 582:1993: Thermal Spraying. Determination of Tensile Adhesive Strength*; European Committee for Standardization: Brussels, Belgium, 1993.
22. *ASTM G65-04(2010): Standard Test Method for Measuring Abrasion Using the Dry Sand/Rubber Wheel Apparatus*; ASTM International: West Conshohocken, PA, USA, 2010.
23. *ASTM G32-10: Standard Test Method for Cavitation Erosion Using Vibratory Apparatus*; ASTM International: West Conshohocken, PA, USA, 2010.
24. Jin, H.W.; Rhyim, Y.M.; Hong, S.G.; Park, C.G. Microstructural evolution of the rapidly quenched Fe–Cr–B alloy thermal spray coatings. *Mater. Sci. Eng. A* **2001**, *304–306*, 1069–1074.
25. Wang, W.H.; Dong, C.; Shek, C.H. Bulk metallic glasses. *Mater. Sci. Eng. R* **2004**, *44*, 45–89.
26. Inoue, A. Glass-forming ability of alloys. *J. Non-Cryst. Solids* **1993**, *156–158*, 473–480.
27. Leffler, B. *STAINLESS—Stainless Steels and Their Properties*, 2nd ed.; Avesta Sheffield AB Research Foundation, Avesta Sheffield Ltd: Sheffield, UK, 1998.
28. Richer, P.; Yandouzi, M.; Beauvais, L.; Jodoin, B. Oxidation behaviour of CoNiCrAlY bond coats produced by plasma, HVOF and cold gas dynamic spraying. *Surf. Coat. Technol.* **2010**, *204*, 3962–3974.
29. Margadant, N.; Neuenschwander, J.; Stauss, S.; Kaps, H.; Kulkarni, A.; Matejicek, J.; Rössler, G. Impact of probing volume from different mechanical measurement methods on elastic properties of thermally

sprayed Ni-based coatings on a mesoscopic scale. *Surf. Coat. Technol.* **2006**, *200*, 2805–2820.
30. Bolelli, G.; Rauch, J.; Cannillo, V.; Killinger, A.; Lusvarghi, L.; Gadow, R. Microstructural and tribological investigation of High-Velocity Suspension Flame Sprayed (HVSFS) Al_2O_3 coatings. *J. Therm. Spray Technol.* **2009**, *18*, 35–49.
31. Zhang, C.; Liu, L.; Chan, K.C.; Chen, Q.; Tang, C.Y. Wear behavior of HVOF-sprayed Fe-based amorphous coatings.*Intermetallics* **2012**, *29*, 80–85.
32. Shebanova, O.N.; Lazor, P. Raman spectroscopic study of magnetite ($FeFe_2O_4$): A new assignment for the vibrational spectrum. *J. Solid State Chem.* **2003**, *174*, 424–430.
33. Varshney, D.; Yogi, A. Structural and transport properties of stoichiometric and Cu^{2+}-doped magnetite: $Fe_{3-x}Cu_xO_4$.*Mater. Chem. Phys.* **2010**, *123*, 434–438.
34. Varshney, D.; Yogi, A. Structural and transport properties of stoichiometric Mn^{2+}-doped magnetite: $Fe_{3-x}Mn_xO_4$.*Mater. Chem. Phys.* **2011**, *128*, 489–494.
35. Uozato, S.; Nakata, K.; Ushio, M. Evaluation of ferrous powder thermal spray coatings on diesel engine cylinder bores. *Surf. Coat. Technol.* **2005**, *200*, 2580–2586.
36. Rabiei, A.; Mumm, D.R.; Hutchinson, J.W.; Schweinfest, R.; Rühle, M.; Evans, A.G. Microstructure, deformation and cracking characteristics of thermal spray ferrous coatings. *Mater. Sci. Eng. A* **1999**, *269*, 152–165.
37. Vetter, J.; Barbezat, G.; Crummenauer, J.; Avissar, J. Surface treatment selections for automotive applications. *Surf. Coat. Technol.* **2005**, *200*, 1962–1968.
38. Holmberg, K.; Matthews, A. *Coating Tribology: Properties, Mechanisms, Techniques and Applications in Surface Engineering*, 2nd ed.; Elsevier: Oxford, UK, 2009.
39. Saikko, V.; Pfaff, H.G. Low wear and friction in alumina/alumina total hip joints: A hip simulator study. *Acta Orthop. Scand.* **1998**, *69*, 443–448.
40. Zhou, Y.S.; Ohashi, M.; Ikeuchi, K. Start up and steady state friction of alumina against alumina. *Wear* **1997**, *210*, 112–119.
41. Milanti, A.; Koivuluoto, H.; Vuoristo, P.; Bolelli, G.; Bozza, F.; Lusvarghi, L. Wear and Corrosion Resistance of High-Velocity Oxygen-Fuel Sprayed Iron-Based Composite Coatings. In Proceedings of ASME 2013 International Mechanical Engineering Congress & Exposition, San Diego, CA, USA, 15–21 November 2013.
42. Valarezo, A.; Sampath, S. An integrated assessment of process-microstructure-property relationships for thermal-sprayed NiCr coatings. *J. Therm. Spray Technol.* **2011**, *20*, 1244–1258.

43. Barletta, M.; Bolelli, G.; Bonferroni, B.; Lusvarghi, L. Wear and corrosion behavior of HVOF-sprayed WC-CoCr coatings on Al alloys. *J. Therm. Spray Technol.* **2010**, *19*, 358–367.

Chapter 10

Design of an Advanced Bearing System for Total Knee Arthroplasty

Mark L. Morrison[1,], Shilesh Jani[2] and Amit Parikh[1]*

[1] *Smith and Nephew Advanced Surgical Devices, 1450 East Brooks Road, Memphis, TN 38116, USA*
[2] *Formerly Smith and Nephew, Currently Orchid Orthopedic Solutions, 4600 East Shelby Drive, Suite 1, Memphis, TN 38118, USA*

ABSTRACT

The objective of this study was to develop an advanced-bearing couple for TKA that optimizes the balance between wear resistance and mechanical properties. The mechanical and structural properties of virgin and highly crosslinked, re-melted UHMWPE were evaluated, and tibial inserts manufactured from these UHMWPE materials were tested against either oxidized zirconium (OxZr) or CoCr femoral components on a knee simulator. This study confirmed that the wear resistance of crosslinked UHMWPE improves with increasing radiation dose but is accompanied by a concomitant reduction in mechanical properties. Compared to CoCr, the ceramic surface of OxZr allows the use of a lower irradiation dose to achieve equivalent reductions in wear rates. As a result, a given wear rate can be achieved without sacrificing the mechanical properties to the same extent that is necessary with a CoCr femoral component. The advantage of ceramic counter bearing surfaces extends to both pristine and microabrasive conditions.

Keywords: highly crosslinked UHMWPE; knee; TKA; wear; mechanical properties; oxidized zirconium; abrasion; crosslink density

1. INTRODUCTION

The most common bearing couple used in total hip and knee arthroplasties (THA and TKA, respectively) is ultra-high molecular weight polyethylene (UHMWPE) articulating against a CoCr alloy. This couple has demonstrated excellent clinical results with survivorship greater than 90% at 13 years in THA [1] and greater than 92% at 13 years in TKA [1]. However, UHMWPE wear and subsequent osteolysis have been a primary long-term failure mechanism, which results in decreasing survivorship with increasing time *in vivo* [2,3,4]. Traditionally, THA and TKA were performed in elderly patients with the intent of reducing pain and providing mobility. However, a number of changes in patient demographics have occurred in the intervening decades, namely: (1) the population has become heavier; (2) patients are receiving THA and TKA at younger ages; and (3) the patients are more likely to continue active lifestyles after surgery [5,6,7]. Therefore, contemporary joint replacements experience greater biomechanical demands than experienced in the past, which create a need for more wear-resistant implants.

This need for improvement was first identified in THA due to the larger number of procedures performed, higher wear rates, and greater incidence of osteolysis. In the late 1990s, crosslinking, which was first used and quizzically abandoned in the 1970s [8], was resurrected as a technique for improved wear resistance of UHMWPE. Today, the formulations of crosslinked UHMWPE that were introduced into the THA marketplace in the late 1990s and early 2000s are showing promising clinical results with up to 12 years of follow-up [9,10,11]. Radiographic wear shows marked improvements, and the incidence of osteolysis secondary to wear shows dramatic reductions compared to conventional UHMWPE.

Historically, highly crosslinked UHMWPE did not see wide-spread use in TKA because crosslinking results in the reduction of particular mechanical properties and contact stresses can be higher in TKA than in THA. However, total knee replacement is now a more prevalent procedure and is predicted to grow at a faster rate than total hip replacement [7]. In addition, Sharkey *et al.* reported that one of the primary, long-term causes of TKA revision was polyethylene wear, which accounted for 25% of the failures in their retrospective review of 212 consecutive TKA revisions [12]. Ten years later, a follow-up study by Sharkey *et al.* [13] reported a remarkable decrease in revisions due to polyethylene wear and concluded that this reduction was, in part, likely due to the successful development of materials with improved wear resistance.

Clearly there is a need for further improvements in TKA bearings technology. Encouraged by the clinical success of crosslinked UHMWPE in the hip, there is now greater confidence that the material may also have a compelling application in knees. Engineers and clinicians, however, recognize that the stresses and motions in knees are different from those in hips and that the effects of the reduction in mechanical properties concomitant with crosslinking must be evaluated carefully. As a result, adoption of crosslinked UHMWPE in TKA has proceeded with greater caution.

It is well understood that the degree of crosslinking appropriate for a particular application is a delicate balance between improvements in wear and reduced mechanical properties resulting from increased crosslinking [14]. With that in mind, the crosslinked UHMWPE formulations currently marketed for TKA typically utilize lower crosslink densities than those marketed for THA.

The hypothesis of the present study is that the wear rates of various formulations of UHMWPE can be reduced not only by increasing crosslink density, but additionally by changing the counter bearing surface from the metallic CoCr alloy to a ceramic. Ceramics (typically sintered aluminum or zirconium oxides) are commonly utilized in THA femoral heads for two reasons, namely (1) reduced wear of UHMWPE because of a lower coefficient of friction compared to CoCr; and (2) improved microabrasion resistance because they are harder than CoCr. Both of these properties are singularly attributed to the predominantly covalently bonded molecular structure of ceramics. However, ceramics are brittle (due to the strong covalent bonds) and have found very limited use in the higher stress application of TKA. A solution to this limitation of bulk ceramics is to design a material system in which the component is fully ceramic only at the bearing surface and metallic in its core. Surface oxidized zirconium (OxZr), trademarked as OXINIUM™ (Smith and Nephew, Memphis, TN, USA), is one such material marketed for orthopaedic implants.

The objective of this study was to develop an advanced-bearing couple for TKA that optimizes the balance between wear resistance and mechanical properties. Wear testing was conducted under pristine and tumbled/abraded conditions to examine the performance and durability of the bearing couples under ideal and less-than-ideal conditions. Mechanical and structural characterizations of various UHMWPEs were also conducted as a function of crosslink density to rationalize the balance between wear resistance and mechanical/structural properties.

2. MATERIALS AND METHODS

2.1. Materials

The UHMWPE materials examined in this study were fabricated from multiple lots of compression-molded GUR1020 UHMWPE rod stock that were obtained from MediTECH Medical Polymers (Fort Wayne, IN, USA). Some of the material was evaluated in the unirradiated state and is denoted as virgin material. Additional rods were crosslinked with nominal gamma-radiation doses of 5.0 Mrad (50 kGy), 7.5 Mrad (75 kGy), or 10.0 Mrad (100 kGy) and subsequently re-melted to stabilize them against oxidation. These highly crosslinked materials are identified as 5-XLPE, 7.5-XLPE and 10-XLPE, respectively. Only the components for wear testing were sterilized by ethylene oxide (EtO) gas. The remainder of the material was evaluated in the unsterilized condition since multiple studies have demonstrated that EtO sterilization does not alter the properties of either virgin or highly crosslinked UHMWPE [15,16,17].

2.2. Mechanical and Physical Characterization

All of the mechanical and physical testing was conducted by MediTECH. Tensile testing ($n = 5$) was performed according to ASTM F648-07 and ASTM D638-03 with Type IV specimens with thicknesses of 3.0 ± 0.05 mm and displacement rates of 50 mm/min. Izod impact testing ($n = 5$) was conducted according to ASTM F648-07 and ASTM D256-05 with a 7.5 J hammer, and density ($n = 3$) was determined according to ASTM D792-00, Method B.

2.3. Thermophysical Analysis

Thermal analysis was conducted according to ASTM F2625-07 using a Netzsch 204 F1 Phoenix differential scanning calorimeter (DSC). Each DSC sample was removed from the core of the rod stock. Samples were cut and weighed to a resolution of 0.01 mg and ranged in mass from approximately 7 to 10 mg. An attempt was made to produce plate-like samples and keep the approximate dimensions consistent in order to minimize dimensional effects on the thermogram variability. The DSC cycle consisted of a 10 min equilibration at 30 °C, followed by heating to 180 °C at 10 °C/min rate, and cooling to 30 °C at 10 °C/min. For all of the thermograms, an empty aluminum crucible was used as the reference. First, an empty aluminum crucible was run as a baseline correction. The samples were crimped into aluminum crucibles and placed in the DSC chamber, which was continuously flushed with research-grade nitrogen gas at a flow rate of approximately 30 mL/min. Five samples ($n = 5$) were run per material condition with this heating profile. The resultant thermograms were analyzed to determine the extrapolated onset (T_{OM}) and peak melting (T_{PM}) temperatures and the heats of fusion (ΔH_m). For the determination of ΔH_m, the integration limits were systematically placed at 50 °C and 160 °C. The percent crystallinity (%X) was estimated as:

$$\%X = \frac{\Delta H_m}{291} \times 100 \quad (1)$$

where 291 J/g is the enthalpy associated with the melting of 100% crystalline polyethylene.

2.4. Molecular Network Parameters

The molecular network parameters were evaluated by Cambridge Polymer Group (Boston, MA, USA). Three cubes (n = 3) with nominal dimensions of 5 × 5 × 5 mm were prepared from each lot of material by Cambridge Polymer. It should be noted that the orientations of the samples relative to the compression axes can affect the results [18] but were unknown in this study. The virgin samples were evaluated according to ASTM D2765-01. The highly crosslinked samples were evaluated according to ASTM F2214-02, and the swell ratio (q_s) and crosslink density (ν_d) were reported. Two lots of virgin UHMWPE, 7 lots of 5-XLPE, 6 lots of 7.5-XLPE, and 3 lots of 10-XLPE were evaluated. The repeat measurements were averaged for each lot of material, and the replicate measurements for each lot were averaged for each level of crosslinking. Briefly, the initial heights (H_o) of the cube specimens were first measured using a digital micrometer with a resolution of 1 μm. The specimens were then immersed in o-xylene with 0.1 wt % Irganox™ antioxidant, and the specimen height was monitored with a lightweight, ceramic dilatometer probe until steady-state conditions were achieved (H_f = final height). As specified in the standard, the swell ratio was calculated according to Equation (2):

$$q_s = \left(\frac{H_f}{H_o}\right)^3 \tag{2}$$

Once q_s was determined, the crosslink density and molecular weight between crosslinks were calculated for all materials according to the following equations delineated in ASTM F2214:

$$\text{Crosslink Density} = \nu_d = \frac{Ln\left(1 - q_s^{-1}\right) + q_s^{-1} + \chi_1 q_s^{-2}}{\phi_1\left(q_s^{-1/3} - \frac{q_s^{-1}}{2}\right)} \tag{3}$$

Molecular Weight Between Crosslinks =

$$M_c = \left(\bar{\nu}\nu_d\right)^{-1} \tag{4}$$

Where:

- χ_1 = Heat of mixing for the polymer-solvent system (Flory interaction parameter) = $0.33 + 0.55q_s^{-1}$
- Φ_1 = Molar volume of the solvent = 136 cm^3/mol
- $\bar{V}$ = Specific volume of the polymer = 920 g/dm^3

2.5. Wear Resistance

The wear resistances of the virgin, 5-XLPE and 7.5-XLPE materials were evaluated on a knee simulator against pristine (*i.e.*, new) CoCr (ASTM F75) and OxZr [19] (ASTM F2384) femoral components. Per standard processing, the cast CoCr femorals were hot isostatic pressed (HIPed) followed by solution heat treating. In addition, testing was conducted with CoCr and OxZr femoral components that were subjected to a tumbling protocol to simulate the effects of microabrasion by third-body debris *in vivo* [20]. This protocol involved tumbling the femoral components in a centrifugal barrel mass-finisher for approximately 30 s in a 25 μm alumina powder and plastic cone media. These tumbled femoral components were then tested on a knee simulator against the various UHMWPE materials to evaluate the effect of the microabrasion on UHMWPE wear.

Genesis II™ (Smith and Nephew, Memphis, TN, USA) cruciate-retaining (CR) knee components ($n = 3$) were tested on a 6-station, displacement-controlled, knee simulator (AMTI, Watertown, MA, USA) and rotated weekly to minimize the effect of station variability. Unloaded soak controls ($n = 3$) were utilized to account for fluid uptake by the UHMWPE tibial inserts. The lubricant was alpha calf fraction (Hyclone Labs, Logan, UT) with 20 mM EDTA and 0.2% sodium azide, which was diluted to 50% with deionized water to achieve an average protein concentration of approximately 20 g/L. These tests were run for approximately 5 million cycles (Mc) each. The lubricant was replaced and the tibial inserts were weighed about every 0.5 Mc. The load/motion profiles were the "high kinematics" inputs described by Barnett *et al.* [21], which was based on a gait-lab study of young, healthy males by Lafortune [22]. In this study, the total A/P translation was 11 mm, which is slightly greater than that used by Barnett *et al.* The slope of the least squares best-fit line of cumulative volume loss *vs.* cycles was determined for each liner, averaged for each bearing combination and defined as the wear rate.

2.6. Roughness Measurements

Roughness measurements of the femoral components were made using a contact profilometer (Surfcom 1800D, Carl Zeiss, Brighton, MI, USA) with a 2 μm radius stylus tip, a 0.25 mm cut-off length and a 4.0 mm evaluation length. Roughness measurements ($n = 10$) were made systematically in the medial-lateral axis from 0° to 45° of flexion on each condyle of each femoral component before wear-simulator testing for both the pristine and tumbled components. Roughness was characterized using five parameters: R_a, the average surface displacement (of peaks and valleys) from the mean surface line; R_{pm}, the average peak height above the mean surface line; R_p, the maximum peak height above the mean surface line; R_{pk}, the average peak height above the mean surface line; and R_{sk}, the skewness, or asymmetry of the profile about the mean surface line.

2.7. Electron Spin Resonance

Evaluation of the residual free-radical concentrations in these materials was conducted to verify the effectiveness of the post-irradiation re-melting process in reducing the free radicals. Samples ($n = 5$) for electron spin resonance (ESR) with diameters of 3 mm and lengths of 10 mm were punched from the center of the rod stock for each material. A 7.5 Mrad gamma-irradiated material without any subsequent heat treatment was utilized as a positive control in this experiment.

The free-radical concentration (FRC) in each material was determined with an X-band ESR spectrometer (Bruker EMX300, University of Memphis) operating at a microwave frequency of approximately 9.8 GHz and employing a high-sensitivity universal X-band resonator cavity. ESR test parameters were constant for all measurements at 1 mW microwave power, 5 Gauss amplitude modulation, 200 G sweep width, 3500 center field, and a relatively high gain and Q-value. The absolute magnitude of the free-radical concentration was analyzed and computed with WinEPR analysis software. The area under the absorption curve of a test specimen was compared to that of the National Institute of Standards and Technology (NIST) intensity standard SRM-2601, and these results were normalized to the specimen's weight.

2.8. Delamination Evaluation

A unidirectional, reciprocating pin-on-disk wear test was conducted to evaluate the susceptibility of the 7.5-XLPE material to delaminate under worst-case conditions. Virgin, unirradiated UHMWPE was used as a negative control in this test because delamination in virgin polyethylene has not been observed clinically. Samples that were gamma sterilized in air were used as a positive control due to the well-known history of *in vivo* delamination for that material [16]. Three replicate samples ($n = 3$) were evaluated per material treatment. Prior to delamination testing, these samples were subjected to accelerated aging according to ASTM F2003-02 for 21 days.

Delamination testing was conducted on a multidirectional pin-on-disk device (OrthoPOD, AMTI, Watertown, MA, USA) with CoCr pins articulating against UHMWPE disks. The CoCr pins were fabricated with a tip radius of 19.81 mm (0.78 in) and polished to a fine, mirror-like finish. A load of 444.8 N (100 lb_f) was applied to each pin to produce an initial, maximum Herztian contact stress of 60.4 MPa, which is at the upper limit of contact stresses reported for TKA *in vivo* [23]. The stroke length was approximately 19 mm. These tests were conducted with reciprocating motion at a frequency of 1.24 Hz for a total of 3 million cycles in 100% alpha calf fraction serum (HyClone, Logan, UT, USA) that was changed at approximately every 500,000 cycles. At those times, the UHMWPE samples were rinsed and examined for signs of delamination with the naked eye. In the case of one gamma-air sample, delamination caused complete wear-through of the UHMWPE and resulted in the removal of this specific sample at less than 3 Mc.

Upon completion of 3 Mc, the samples were removed and any signs of delamination were noted. The samples were then sectioned along the long axis of the wear track, cut with a microtome to produce a relatively smooth cross-section, sputter-coated with gold, and examined with a scanning electron microscope (SEM) in secondary electron mode. Both the wear surface and the sub-surface were examined for any signs of cracking or delamination.

2.9. Statistical Analyses

Statistical analyses of the results were performed with Minitab 16 (Minitab, Inc., State College, PA, USA). The homogeneity of variances within the various data sets was evaluated with Levene's test. If the variances were homogeneous, either the Student's t-test with pooled variances or analysis of variance (ANOVA) with Tukey's *post hoc* test was conducted to determine statistically significant differences in the properties. If the variances were not homogeneous, either the Student's t-test with separate variances or Welch analysis of variance (ANOVA) with a Tukey's *post hoc* test was conducted to determine statistically significant differences in the properties. The level of significance (α) was 0.05.

3. RESULTS AND DISCUSSION

3.1. Mechanical and Physical Characterization

While the yield strength (YS) exhibited small, but significant decreases ($p < 0.001$) of approximately 8%–9% upon irradiation to any dose (Table 1), the ultimate tensile strength (UTS) and elongation at break decreased linearly with increasing radiation dose with statistically significant differences between each dose (Figure 1). Compared to virgin, the UTS decreased 10%–21% ($p < 0.001$) and elongation decreased 27%–45% ($p < 0.001$) depending upon the selected irradiation dose (Figure 1b,c). Likewise, the impact strength decreased approximately 45%–59% upon crosslinking with statistically significant differences ($p < 0.001$) between each dose (Figure 1d).

Table 1. Summary of the mean (±standard deviations) properties evaluated in this study for compression-molded GUR1020 UHMWPE in the virgin and highly crosslinked, re-melted conditions.

Material Property	Nominal Radiation Dose (Mrad)			
	0.0	5.0	7.5	10.0
Yield Strength, YS (MPa)	22.6 ± 0.4	20.7 ± 0.3	20.8 ± 0.5	20.6 ± 0.2
Ultimate Tensile Strength, UTS (MPa)	54.9 ± 1.4	49.4 ± 2.3	46.9 ± 1.8	43.6 ± 2.5
Elongation at Break, EL (%)	516 ± 29	379 ± 16	323 ± 8	282 ± 15
Izod Impact Strength (kJ/m^2)	164 ± 13	90 ± 4	76 ± 2	67 ± 3
Density (g/cm^3)	0.936 ± 0.001	0.932 ± 0.001	0.932 ± 0.001	0.932 ± 0.002
Onset Melting Temperature, T_{OM} (°C)	125.9 ± 0.5	123.6 ± 0.4	122.2 ± 0.6	120.9 ± 1.0
Peak Melting Temperature, T_{PM} (°C)	140.8 ± 0.6	139.8 ± 0.4	141.2 ± 0.7	141.6 ± 0.4
Crystallinity (%)	55.9 ± 0.8	50.6 ± 0.6	50.7 ± 1.2	50.7 ± 0.9
Swell Ratio, q	17.57 ± 0.26	3.57 ± 0.10	3.08 ± 0.10	2.82 ± 0.17
Extract (%)	18.5 ± 3.5	--	--	--
Crosslink Density, v_d (mol/dm^3)	0.011 ± 0.003	0.151 ± 0.008	0.204 ± 0.024	0.237 ± 0.018
Molecular Weight Between Crosslinks, M_c (10^3 g/mol)	90.55 ± 16.39	6.11 ± 0.31	4.62 ± 0.54	3.94 ± 0.30
Free-Radical Concentration, FRC (spins/g)	Not Detectable	Not Detectable	Not Detectable	Not Detectable
Maximum Oxidation Index, OI_{max}	0.02 ± 0.00	0.02 ± 0.01	0.01 ± 0.00	--

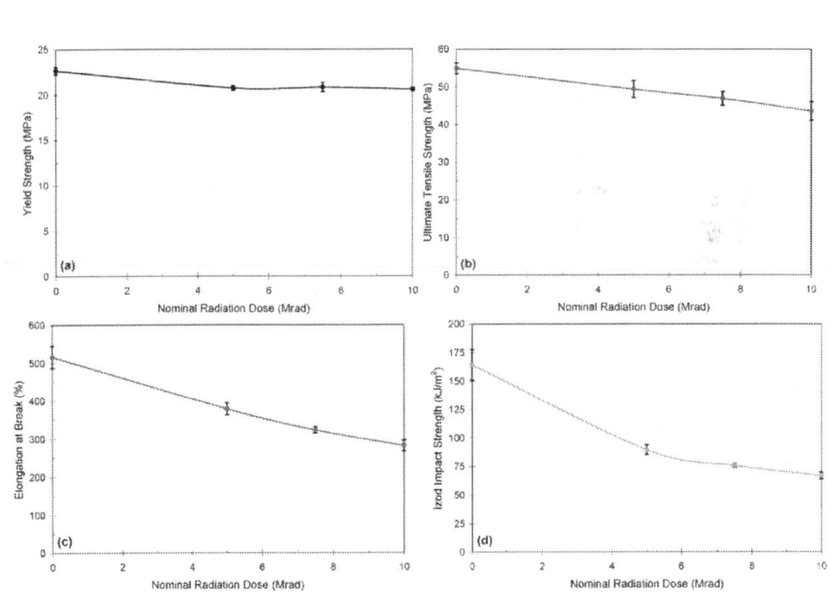

Figure 1. Plots of the (**a**) yield strength; (**b**) ultimate tensile strength; (**c**) elongation at break; and (**d**) Izod impact strength for GUR1020 UHMWPE irradiated with various gamma-radiation doses and subsequently re-melted.

3.2. Thermophysical Analysis

As demonstrated in Table 1, the T_{OM} was found to decrease by up to 4% in the crosslinked materials compared to the virgin materials with statistically significant differences ($p < 0.001$) for virgin compared to 7.5-XLPE and 10-XLPE and for 5-XLPE compared to 10-XLPE. The T_{PM} changed by ±1% or less in the crosslinked materials, relative to the virgin material, and some of these differences were statistically significant ($p < 0.001$). Finally, the %X decreased by approximately 10%, compared to virgin, upon irradiation to any of the irradiation doses examined in this study ($p < 0.001$).

3.3. Molecular Network Parameters

The mean swell ratio, percent extract, crosslink density, and molecular weight between crosslinks for the virgin and highly crosslinked materials are summarized in Table 1. The swell ratio (Figure 2) and molecular weight between crosslinks decreased with increasing radiation dose, and the crosslink density (Figure 3) increased with increasing radiation dose. The changes in all of these metrics were statistically significant ($p < 0.001$) as the dose increased to 7.5 Mrad. The differences in these metrics between the 7.5-XLPE and 10-XLPE materials was not significant at the levels evaluated in this study. Within the dose range examined, these values were similar to those reported by Greer and co-authors [24] as demonstrated in Figure 2 and Figure 3.

3.4. Roughness Measurements

The mean roughness parameters in the pristine conditions met the applicable quality specifications and standards for both femoral materials (Table 2 and Table 3). After the CoCr femoral components were subjected to the tumbling protocol, all of the roughness parameters evaluated in this study increased ($p \leq 0.016$, Table 2). For the OxZr femorals (Table 3), all of the roughness parameters except for R_a ($p = 0.062$) and R_{sk} increased by statistically significant amounts ($p \leq 0.031$) after tumbling. In contrast, R_{sk} decreased significantly ($p < 0.001$) after tumbling. Comparing the materials in the tumbled conditions, all of the roughness parameters were significantly higher for CoCr relative to OxZr ($p < 0.001$).

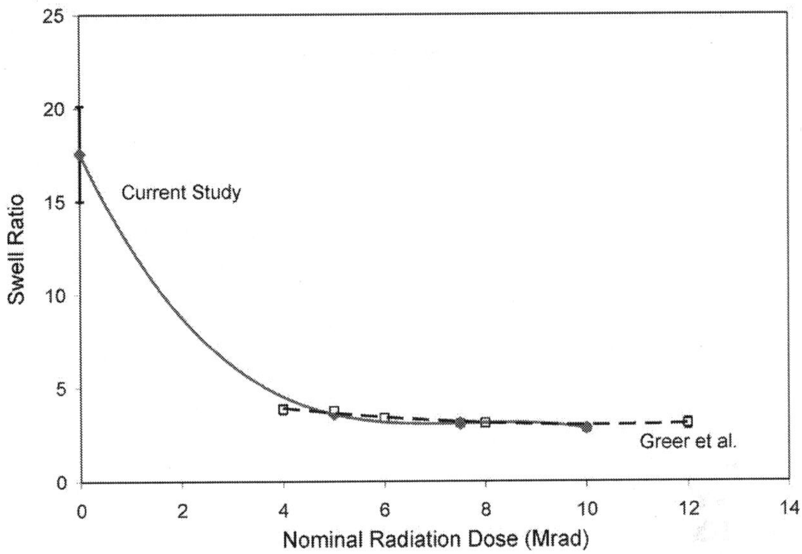

Figure 2. Plot of swell ratio for compression-molded GUR1020 UHMWPE in the virgin state and gamma-irradiated to various doses and re-melted. The results reported by Greer *et al.* [24] are also plotted for comparison. The error bars represent the standard deviations.

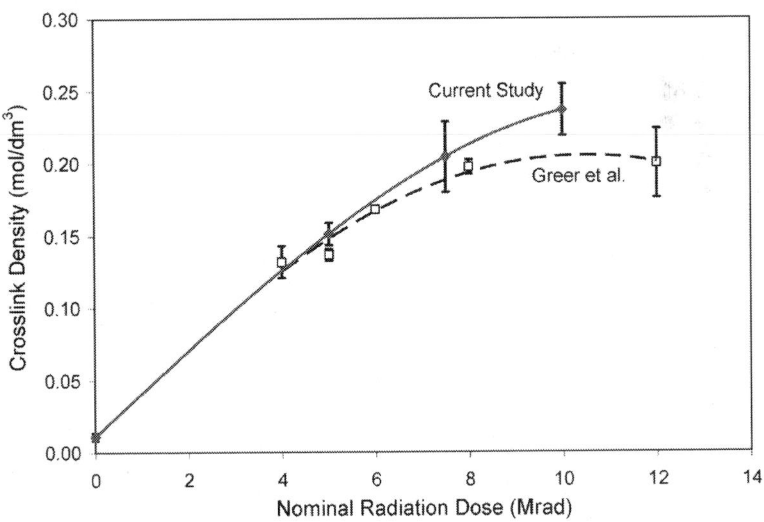

Figure 3. Plot of crosslink density for compression-molded GUR1020 UHMWPE in the virgin state and gamma-irradiated to various doses and re-melted. The results reported by Greer *et al.* [24] are also plotted for comparison. The error bars represent the standard deviations.

Table 2. Summary of the mean (± standard deviations) roughness parameters for the CoCr femoral components in the pristine and roughened conditions. The levels of significance (p) for the differences between the pristine and roughened components are also reported.

Roughness Parameter	Pristine CoCr	Roughened CoCr	p
R_a (μm)	0.035 ± 0.004	0.077 ± 0.009	<0.001
R_{pm} (μm)	0.140 ± 0.013	0.510 ± 0.057	<0.001
R_p (μm)	0.216 ± 0.028	1.658 ± 0.160	<0.001
R_{pk} (μm)	0.066 ± 0.008	0.253 ± 0.022	<0.001
R_{sk}	0.618 ± 0.115	2.008 ± 1.363	0.016

Table 3. Summary of the mean (± standard deviations) roughness parameters for the OxZr femoral components in the pristine and roughened conditions. The levels of significance (p) for the differences between the pristine and roughened components are also reported.

Roughness Parameter	Pristine OxZr	Roughened OxZr	p
R_a (μm)	0.044 ± 0.007	0.050 ± 0.002	0.062
R_{pm} (μm)	0.152 ± 0.027	0.179 ± 0.011	0.001
R_p (μm)	0.248 ± 0.088	0.397 ± 0.066	0.019
R_{pk} (μm)	0.052 ± 0.009	0.072 ± 0.006	<0.001
R_{sk}	−1.443 ± 0.683	−2.801 ± 0.557	<0.001

3.5. Wear Resistance

Wear testing of pristine CoCr femoral components against virgin and highly crosslinked UHMWPE tibial inserts reflected documented trends in the literature [14] in that the mean wear rate decreased significantly ($p < 0.001$) with each increase in the radiation dose examined in this study (Figure 4). The mean wear rates (± standard deviations) of the virgin, 5-XLPE, and 7.5-XLPE tibial inserts were 23.4 ± 2.4 mm^3/Mc, 10.9 ± 0.9 mm^3/Mc, and 6.4 ± 0.6 mm^3/Mc, respectively. As a result, 5-XLPE produced a 53% reduction in wear while the use of 7.5-XLPE resulted in a 73% reduction in wear compared to virgin UHMWPE. Based on this trend with pristine CoCr femoral components, it is estimated that a nominal radiation dose of about 10 Mrad is necessary to produce a wear rate that is not measurable by the gravimetric technique.

Knee simulator wear testing of pristine OxZr femoral components against virgin, 5-XLPE, and 7.5-XLPE tibial inserts yielded wear rates of 11.7 ± 1.9 mm^3/Mc, 3.1 ± 1.0 mm^3/Mc, and 1.4 ± 0.2 mm^3/Mc, respectively. As a result, 5-XLPE produced a 73% reduction in wear while the use of 7.5-XLPE resulted in an 88% reduction in wear compared to virgin UHMWPE against pristine OxZr femoral components. Statistical analysis demonstrated that the wear rates for the highly crosslinked materials were significantly lower than that of the virgin material ($p < 0.001$), but there was insufficient evidence of differences between the 5-XLPE and 7.5-XLPE at the levels tested in this study. This data suggests that an estimated, nominal radiation dose of about 8 Mrad is necessary to produce a wear rate that is not measurable by the gravimetric technique when paired with pristine OxZr (Figure 4). More importantly, pristine OxZr femoral components alone reduced wear by 50% to 79% ($p \leq 0.001$) compared to pristine CoCr femoral components tested against the same UHMWPE materials.

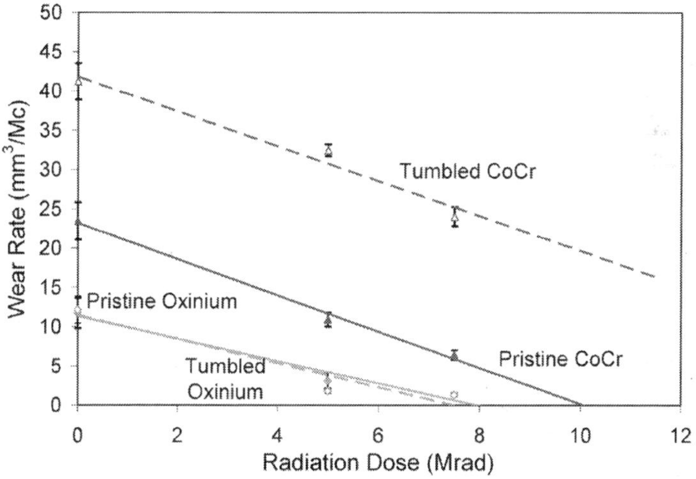

Figure 4. Plot of the mean wear rates (±standard deviations) in a knee simulator for UHMWPE crosslinked to various doses against either CoCr or OxZr femoral components in pristine (solid symbols and lines) and tumbled (open symbols and dashed lines) conditions. The regression lines have been extrapolated to radiation doses greater than 7.5 Mrad.

When the CoCr femoral components were tumbled to simulate *in vivo* microabrasion of the surfaces due to third-body debris, the wear rates for each material increased by statistically significant amounts ($p \leq 0.001$) compared to those calculated against pristine CoCr femoral components (Figure 4). In these tests, the mean wear rates (± standard deviations) of virgin, 5-XLPE, and 7.5-XLPE tibial inserts were 41.2 ± 2.3 mm^3/Mc, 32.5 ± 0.7 mm^3/Mc, and 24.0 ± 1.2 mm^3/Mc, respectively. This data demonstrates that, although the trend between radiation dose and wear rate is similar, the dramatic improvements in wear resistance for highly crosslinked materials are degraded by abrasion of the CoCr femoral components.

In contrast, no statistically significant changes in wear rates ($p \geq 0.113$) were observed, relative to pristine OxZr, when each material was tested against OxZr femoral components that were exposed to the same microabrasive challenge (Figure 4). For these simulator tests, the mean wear rates (±standard deviations) of virgin, 5-XLPE, and 7.5-XLPE tibial inserts were 12.2 ± 1.7 mm^3/Mc, 1.8 ± 0.4 mm^3/Mc, and 1.3 ± 0.1 mm^3/Mc, respectively. Under these microabrasive conditions, the wear rates against tumbled OxZr remained 49% to 83% lower ($p \leq 0.003$) than the wear rates observed for pristine CoCr femoral components tested against the same UHMWPE materials.

3.6. Electron Spin Resonance

The virgin, 5-XLPE, 7.5-XLPE, and 10-XLPE specimens did not show detectable ESR signals, indicating that the free-radical concentration (Table 1) was below the sensitivity of the instrument, which is approximately 1×10^{13} spins/g. In contrast, the 7.5 Mrad irradiated, unannealed material (positive control) exhibited a high mean FRC (±SD) of $2.36 \times 10^{16} \pm 0.36 \times 10^{16}$ spins/g, which is 3 orders of magnitude greater than the instrument sensitivity.

3.7. Delamination Study

The virgin and highly crosslinked samples endured 3 Mc of testing without any evidence of cracking or delamination on the wear surfaces (Figure 5a,b) or on the longitudinal cross-sections of the wear tracks in the post-test examinations. In contrast, all of the gamma-air-sterilized samples (Figure 5c) exhibited the onset of delamination prior to the completion of the full 3 Mc test protocol.

In this pin-on-disk study, the highly crosslinked UHMWPE materials did not exhibit a propensity for delamination under worst-case testing conditions. In addition, this particular test protocol was validated by the absence of delamination in the negative control and mild to extensive delamination in the positive controls, which replicated the damage observed in clinical retrievals. The resistance of highly crosslinked and re-melted UHMWPE to delamination has been similarly demonstrated by others in both pin-on-disk [25] and knee simulator testing [26].

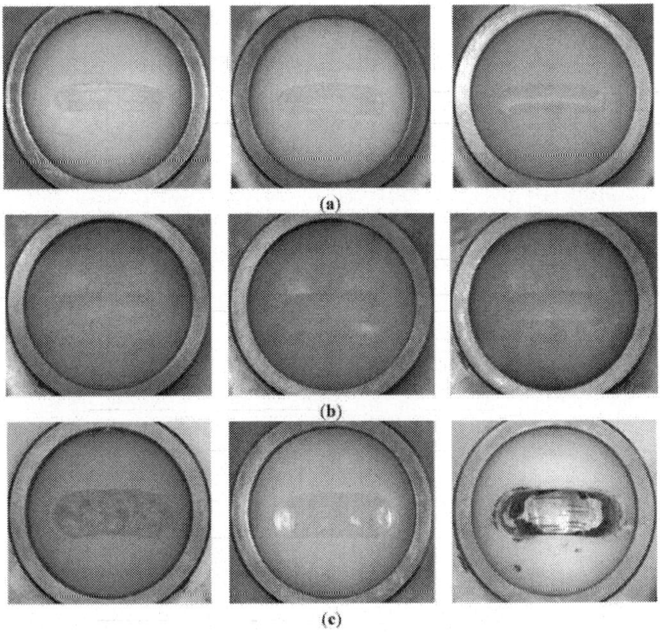

Figure 5. Photomacrographs of the (**a**) virgin; (**b**) 7.5-XLPE; and (**c**) gamma-air-sterilized samples after 3 million cycles of reciprocating pin-on-disk testing. Although wear tracks are visible on the virgin and 7.5-XLPE samples, no cracking or delamination was observed.

4. DISCUSSION

This study confirmed that increasing crosslink density (*i.e.*, increasing the radiation dose) results in significant reductions in wear rates. This trend was independent of the femoral bearing material. When the femoral component was pristine CoCr alloy, the wear rate *versus* radiation dose trend suggests that a crosslink dose of 10 Mrad would result in wear rates that are not measurable by the gravimetric technique. This does not mean that the wear rate is zero, rather that the wear rate is low enough that gravimetric measurement errors cannot distinguish it from zero as has been previously reported for hip simulator testing on 10 Mrad irradiated and re-melted UHMWPE [27].

This study also confirmed that the ceramic bearing surface of oxidized zirconium resulted in greater reductions in wear of crosslinked UHMWPE than a pristine CoCr bearing surface. When the relationships between wear rate and radiation dose for pristine CoCr and ceramic oxidized zirconium are compared, it was found that oxidized zirconium was more effective in reducing the wear of UHMWPE at a given radiation dose. This finding was expected considering the widely documented advantage of ceramic bearing materials when tested against UHMWPE in both THA [28,29] and TKA [30,31,32]. Subsequent knee-simulator testing has demonstrated that the low wear provided by OxZr femorals paired with 7.5-XLPE tibial inserts is maintained for an unprecedented 45 Mc of simulator testing [33,34].

The factorial (or percent) differences in UHMWPE wear rates between CoCr and OxZr increase as the crosslinking dose increases; 50% for virgin UHMWPE and 79% for 7.5 Mrad UHMWPE. However, the absolute difference in wear rates between CoCr and OxZr diminishes as the crosslink dose increases; 11.7 mm^3/Mc for virgin *vs.* 5 mm^3/Mc for 7.5 Mrad UHMWPE. The ramification of these trends is that a lower crosslinking dose can be used with OxZr than with CoCr to achieve a given wear reduction. For instance, in order to achieve a wear rate of approximately 3 mm^3/Mc, one would need to use a crosslinking dose of about 6 Mrad with OxZr *vs.* approximately 9 Mrad with CoCr. This additional 3 Mrad of radiation has consequences to the mechanical and structural properties of UHMWPE, as will be discussed later.

The discussion thus far has focused on knee simulator wear testing of femoral components in the pristine condition. When the femoral components were subjected to a microabrasive tumbling protocol, as expected the metallic, CoCr surfaces roughen to a greater extent than the harder [35], ceramic surfaces of OxZr. After tumbling, the roughness metrics measured on the CoCr components were well within the range reported previously for clinical retrievals [36]. Because the ceramic surface of OxZr is twice as hard as CoCr (12.1 GPa and 5.5 GPa, respectively, based on nanoindentation per the Oliver-Pharr method [35]), it was expected to roughen to a lesser degree than CoCr. However, the finding that the wear rates of UHMWPE against OxZr were unaffected by the tumbling protocol was unexpected, because OxZr did, indeed, exhibit increases in the roughness parameters. Understanding this observation requires a more detailed review of the roughness data.

The tumbling of CoCr and OxZr resulted in increases in the average surface roughness (R_a) for both materials. However, the increase in R_a was greater for CoCr (+122%, $p < 0.001$) than for OxZr (+12%, $p = 0.062$) relative to the respective pristine materials. Similar trends were observed for R_p, R_{pm}, and R_{pk}. The surface skewness (R_{sk}) exhibited a different behavior and illustrates the primary difference between metallic and ceramic counterfaces. CoCr femorals almost always exhibited positive values of R_{sk}, which means that there were more peaks than valleys on the bearing surface. After the CoCr femorals were subjected to the tumbling protocol, the scratches resulted in a greater quantity of larger peaks due to pile-up on the sides of the scratches, and the R_{sk} generally increased to greater positive values (+225%, $p = 0.016$). In contrast, the OxZr femorals always exhibited negative values of R_{sk}, which means that there are more valleys than peaks. After the tumbling protocol, R_{sk} decreased (−94%, $p < 0.001$) to more negative values due to scratching of the surface (*i.e.*, valleys) with little pile-up (*i.e.*, peaks), which is characteristic of ceramic bearing surfaces [37].

The wear rates of each of the UHMWPE formulations increased when CoCr was roughened compared to the corresponding wear rates under pristine conditions. During articulation of these components, the hard peaks on the CoCr femorals plow into the softer UHMWPE tibial inserts and generate greater amounts of wear. When UHMWPE was tested against OxZr subjected to the same microabrasive challenge, the resulting knee simulator wear rates were not elevated compared to the corresponding wear rates under pristine conditions. This lack of increase in wear after tumbling is primarily due to the smoother counterface with negative skewness. In other words, the tumbled OxZr femorals do not have many peaks that can plow into the UHMWPE tibial inserts and generate wear debris. As a result, wear does not increase even after a microabrasive challenge.

When these tumbled components were tested in a knee simulator against virgin and the various formulations of highly crosslinked UHMWPEs, the trends in wear rates were similar to those noted above for pristine femoral components. Increasing the crosslink dose resulted in improved wear resistance against both femoral materials. However, the wear rates increased when CoCr was abraded, while they did not change when OxZr was tumbled. If one wanted to achieve a wear rate of approximately 3 mm^3/Mc with tumbled OxZr as the counter bearing surface, a crosslink dose of about 6 Mrad would be required. To achieve that same 3 mm^3/Mc wear rate with abraded CoCr as the counter bearing surface, a predicted crosslink dose of approximately 18 Mrad would be required. Such a high crosslink dose is unlikely to be useful clinically because of the attendant loss of mechanical properties.

The wear-test results discussed above have highlighted the following trends: (1) the wear rates decrease with increasing radiation dose (crosslink density); (2) this trend is similar for both of the femoral bearing materials evaluated; (3) this trend is independent of the surface roughness of the counter bearing surface; (4) ceramic bearing surfaces result in lower UHMWPE wear rates than CoCr alloy bearing surfaces independent of the crosslinking dose; (5) wear rates of all formulations of UHMWPE increase markedly when CoCr counter bearing surfaces are abraded; and (6) wear rates of all formulations of UHMWPE are unaffected when OxZr ceramic counter bearing surfaces are subjected to a microabrasive challenge.

This study also confirmed previous studies [24,38,39] that showed that mechanical properties are affected by radiation crosslinking, particularly ultimate tensile strength (UTS), elongation at break, and impact strength. It stands to reason that the mechanically strongest material would be the ideal choice for use in orthopaedic implants. However, the mechanically strongest material (virgin, non-crosslinked UHMWPE) is also the least resistant to wear. Therefore, a balance between wear resistance and mechanical properties needs to be achieved and should also take into account the design of the orthopaedic implant. As discussed above, the ability to use a ceramic counter bearing surface affords greater reduction in wear rates with a given level of crosslinking as compared to CoCr surfaces. This study demonstrated that ceramic surfaces provide reductions in wear that are equivalent to an additional 3–5 Mrad crosslinking dose compared to pristine CoCr. When CoCr is roughened, this advantage of ceramics is elevated by an estimated 12 Mrad of crosslinking dose.

The crosslinked UHMWPE materials tested in this study were all re-melted after irradiation, and did not undergo any further irradiation. Because of this thermal treatment, all crosslinked formulations were found to have no measurable free radicals, which improves the resistance to oxidation [40]. However, it should be noted that UHMWPE that is prone to oxidation is also prone to *in vivo* degradation, particularly subsurface delamination wear [16]. This study confirmed that virgin (non-irradiated) and crosslinked, re-melted UHMWPEs are resistant to delamination wear after accelerated aging. Under identical conditions, the UHMWPE material that was irradiated in air, and subsequently aged showed copious delamination as has been documented clinically.

In summary, this study confirmed that crosslinking of UHMWPE for improved wear resistance comes at the expense of reduced mechanical properties. For applications in TKA, a majority of the research to-date has focused on striking the right balance between wear resistance and mechanical properties through selection of the irradiation dose alone. This study, however, showed that the counter bearing surface can have a profound effect on the wear rates of UHMWPEs. Compared to CoCr, the ceramic surface of OxZr allows the use of a lower irradiation dose to achieve equivalent reductions in wear rates. As a result, a given wear rate can be achieved without sacrificing the mechanical properties to the same extent that is necessary if a CoCr femoral component is used. The advantage of ceramic counter bearing surfaces extends to both pristine and microabrasive conditions.

5. CONCLUSIONS

This study confirmed that the wear resistance of crosslinked UHMWPE in TKA improves with increasing irradiation dose. However, along with this advantage is the concomitant reduction in key mechanical properties in highly crosslinked UHMWPE. When CoCr is used as the counter bearing surface, the appropriate balance between the wear resistance and the mechanical properties of crosslinked UHMWPE can be controlled only by the choice of the irradiation dose within a narrow range. On the other hand, when a ceramic counter bearing surface (e.g., oxidized zirconium) is used in the bearing couple under ideal, pristine conditions, the resultant improvement in wear resistance is equivalent to an additional 3–5 Mrad of radiation without sacrificing additional UHMWPE mechanical properties. This improvement in wear resistance is expanded even further to the equivalent of an approximate dose of 12 Mrad under microabrasive conditions.

The oxidized zirconium ceramic bearing surface articulating against UHMWPE crosslinked to a dose of 7.5 Mrad can provide a wear resistance that is similar to pristine CoCr articulating against approximately 10 Mrad crosslinked UHMWPE. If microabrasion of the CoCr femoral occurs *in vivo*, this difference in wear resistance grows even greater. Furthermore, the mechanical properties of the UHMWPE are preserved due to the lower irradiation dose that is necessary to achieve a given wear rate against oxidized zirconium.

Finally, re-melting of the radiation crosslinked UHMWPE results in elimination of all measurable free radicals and imparts oxidative stability to the material and attendant resistance to delamination wear.

ACKNOWLEDGMENTS

Gary Hines and Carolyn Weaver are gratefully acknowledged for their assistance in this study.

AUTHOR CONTRIBUTIONS

Mark Morrison, Shilesh Jani and Amit Parikh conceived and designed the experiments; Mark Morrison and Amit Parikh performed the experiments and analyzed the data; Mark Morrison, Shilesh Jani and Amit Parikh wrote the paper.

REFERENCES

1. Australian Orthopaedic Association National Joint Replacement Registry. *Annual Report*; Australian Orthopaedic Association: Adelaide, Australia, 2014.

2. Archibeck, M.J.; Jacobs, J.J.; Roebuck, K.A.; Glant, T.T. The basic science of periprosthetic osteolysis. *J. Bone Jt. Surg. Am.* **2001**, *82*, 1478–1489.
3. Harris, W.H. Wear and periprosthetic osteolysis: The problem. *Clin. Orthop. Relat. Res.* **2001**, *393*, 66–70.
4. Gupta, S.K.; Chu, A.; Ranawat, A.S.; Slamin, J.; Ranawat, C.S. Review article: Osteolysis after total knee arthroplasty.*J. Arthroplast.* **2007**, *22*, 787–799.
5. D'Apuzzo, M.R.; Hernandez-Polo, V.H.; Sierra, R.J. *National Trends in Primary Total Knee Arthroplasty: A Population-Based Study*; American Academy of Orthopaedic Surgeons: New Orleans, LA, USA, 2010; p. 681.
6. Dahl, A.W.; Robertsson, O.; Lidgren, L. *Surgical Treatment for Knee OA in Younger Patients*; American Academy of Orthopaedic Surgeons: New Orleans, LA, USA, 2010; p. 126.
7. Crowninshield, R.D.; Rosenberg, A.G.; Sporer, S.M. Changing demographics of patients with total joint replacement.*Clin. Orthop. Relat. Res.* **2006**, *443*, 266–272.
8. Kurtz, S.M.; Ong, K. Contemporary total hip arthroplasty: Hard-on-hard bearings and highly crosslinked UHMWPE. In *UHMWPE Biomaterials Handbook*, 2nd ed.; Kurtz, S.M., Ed.; Academic Press: Burlington, MA, USA, 2009; pp. 55–79.
9. Epinette, J.A.; Manley, M.T. No differences found in bearing related hip survivorship at 10–12 years follow-up between patients with ceramic on highly cross-linked polyethylene bearings compared to patients with ceramic on ceramic bearings. *J. Arthroplast.* **2014**, *29*, 1369–1372.
10. Glyn-Jones, S.; Thomas, G.E.R.; Garfjeld-Roberts, P.; Gundle, R.; Taylor, A.; McLardy-Smith, P.; Murray, D.W. The John Charnley award: Highly crosslinked polyethylene in total hip arthroplasty decreases long-term wear: A double-blind randomized trial. *Clin. Orthop. Relat. Res.* **2015**, *473*, 432–438.
11. Engh, C.A., Jr.; Hopper, R.H., Jr.; Huynh, C.; Ho, H.; Sritulanondha, S.; Engh, C.A., Sr. A prospective, randomized study of cross-linked and non–cross-linked polyethylene for total hip arthroplasty at 10-year follow-up. *J. Arthroplast.***2012**, *27*, 2–7.
12. Sharkey, P.F.; Hozack, W.J.; Rothman, R.H.; Shastri, S.; Jacoby, S.M. Insall award paper. Why are total knee arthroplasties failing today? *Clin. Orthop. Relat. Res.* **2002**, *404*, 7–13.
13. Sharkey, P.F.; Lichstein, P.M.; Shen, C.; Tokarski, A.T.; Parvizi, J. Why are total knee arthroplasties failing today—Has anything changed after 10 years? *J. Arthroplast.* **2014**, *29*, 1774–1778.

14. Asano, T.; Akagi, M.; Clarke, I.C.; Masuda, S.; Ishii, T.; Nakamura, T. Dose effects of cross-linking polyethylene for total knee arthroplasty on wear performance and mechanical properties. *J. Biomed. Mater. Res. B* **2007**, *83*, 615–622.
15. Muratoglu, O.K.; Merrill, E.W.; Bragdon, C.R.; O'Connor, D.; Hoeffel, D.; Burroughs, B.; Jasty, M.; Harris, W.H. Effect of radiation, heat, and aging on *in vitro* wear resistance of polyethylene. *Clin. Orthop. Relat. Res.* **2003**, *417*, 253–262.
16. Williams, I.R.; Mayor, M.B.; Collier, J.P. The impact of sterilization method on wear in knee arthroplasty. *Clin. Orthop. Relat. Res.* **1998**, *356*, 170–180.
17. Goldman, M.; Lee, M.; Pruitt, L.; Gronsky, R. *A Comparison of Sterilization Techniques on the Structure and Morphology of Medical Grade UHMWPE*; Society for Biomaterials: Toronto, ON, Canada, 1996; p. 189.
18. Terrill, L.; Liao, C.W.; Lin, S.; DeAzevedo, M.; Sun, D.C. *Effects of Anisotropy in UHMWPE on Crosslinked Density Testing*; Orthopaedic Research Society: San Francisco, CA, USA, 2008; p. 1674.
19. Sheth, N.P.; Lementowski, P.; Hunter, G.; Garino, J.P. Clinical applications of oxidized zirconium. *J. Surg. Orthop. Adv.***2008**, *17*, 17–26.
20. Widding, K.; Hines, G.; Hunter, G.; Salehi, A. *Knee Simulator Protocol for Testing of Oxidized Zirconium and Cobalt Chrome Femoral Components under Abrasive Conditions*; Orthopaedic Research Society: Dallas, TX, USA, 2002; p. 1009.
21. Barnett, P.I.; Fisher, J.; Auger, D.D.; Stone, M.H.; Ingham, E. Comparison of wear in a total knee replacement under different kinematic conditions. *J. Mater. Sci. Mater. Med.* **2001**, *12*, 1039–1042.
22. Lafortune, M.A.; Cavanagh, P.R.; Sommer, H.J., III; Kalenak, A. Three-dimensional kinematics of the human knee during walking. *J. Biomech.* **1992**, *25*, 347–357.
23. Bartel, D.L.; Rawlinson, J.J.; Burstein, A.H.; Ranawat, C.S.; Flynn, W.F., Jr. Stresses in polyethylene components of contemporary total knee replacements. *Clin. Orthop. Relat. Res.* **1995**, *317*, 76–82.
24. Greer, K.W.; King, R.S.; Chan, F.W. The effects of raw material, irradiation dose, and irradiation source on crosslinking of UHMWPE. In *Crosslinked and Thermally Treated Ultra-High Molecular Weight Polyethylene for Joint Replacements, ASTM STP 1445*; Kurtz, S.M., Gsell, R., Martell, J., Eds.; ASTM International: West Conshohocken, PA, USA, 2003; pp. 209–220.

25. Burroughs, B.; Muratoglu, O.K.; O'Connor, D.O.; Bragdon, C.R.; Harris, W.H. *Development of a Pitting/Delamination Pin on Disc Wear Model to Simulate the Failure Modes Often Seen in UHMWPE Tibial Knee Inserts*; Orthopaedic Research Society: Orlando, FL, USA, 2000; p. 0552.
26. Deluzio, K.J.; Muratoglu, O.K.; O'Connor, D.O.; Bragdon, C.R.; O'Flynn, H.; Rubash, H.; Jasty, M.; Wyss, U.P.; Harris, W.H. *Development of an in vitro Knee Delamination Model in a Knee Simulator with Physiologic Load and Motion*; World Biomater Cong: Kamuela, HI, USA, 2000; p. 842.
27. Good, V.; Ries, M.; Barrack, R.L.; Widding, K.; Hunter, G.; Heuer, D. Reduced wear with oxidized zirconium femoral heads. *J. Bone Jt. Surg. Am.* **2003**, *85A*, 105–110.
28. Dahl, J.; Nivbrant, B.; Soderlund, P.; Nordsletten, L.; Rohrl, S.M. Less wear with aluminumoxide heads against conventional polyethylene—A 10 year follow-up radiostereometry (RSA). *J. Bone Jt. Surg. Br.* **2010**, *92*, 392–397.
29. Wroblewski, B.M.; Siney, P.D.; Dowson, D.; Collins, S.N. Prospective clinical and joint simulator studies of a new total hip arthroplasty using alumina ceramic heads and cross-linked polyethylene cups. *J. Bone Jt. Surg. Br.* **1996**, *78*, 280–285.
30. Kaddick, C.; Streicher, R. Ceramic total knee replacements: Do they produce less wear? *Bone Jt. J.* **2013**, *95*, 94–97.
31. Oonishi, H.; Ueno, M.; Kim, S.C.; Oonishi, H.; Iwamoto, M.; Kyomoto, M. Ceramic *vs.* cobalt-chrome femoral components; wear of polyethylene insert in total knee prosthesis. *J. Arthroplast.* **2009**, *24*, 374–382.
32. Zietz, C.; Bergschmidt, P.; Lange, R.; Mittelmeier, W.; Bader, R. Third-body abrasive wear of tibial polyethylene inserts combined with metallic and ceramic femoral components in a knee simulator study. *Int. J. Artif. Org.* **2013**, *36*, 47–55.
33. Papannagari, R.; Hines, G.; Sprague, J.; Morrison, M. *Long-Term Wear Performance of an Advanced Bearing Knee Technology*; ISTA: Dubai, UAE, 2010; pp. A12–A13.
34. Papannagari, R.; Hines, G.; Morrison, M.; Sprague, J. *Long-Term Wear Performance of an Advanced Bearing Technology for TKA*; Orthopaedic Research Society: Long Beach, CA, USA, 2011; p. 1141.
35. Long, M.; Riester, L.; Hunter, G. *Nano-Hardness Measurements of Oxidized Zr-2.5Nb and Various Orthopaedic Materials*; Society Biomater: San Diego, CA, USA, 1998; p. 528.
36. Levesque, M.; Livingston, B.J.; Jones, W.M.; Spector, M. *Scratches on Condyles in Normal Functioning Total Knee Arthroplasty*; Orthopaedic Research Society: New Orleans, LA, USA, 1998; pp. 241–247.

37. Cooper, J.R.; Dowson, D.; Fisher, J.; Jobbins, B. Ceramic bearing surfaces in total artificial joints: Resistance to third body wear damage from bone cement particles. *J. Med. Eng. Technol.* **1991**, *15*, 63–67.
38. Pruitt, L.A. Deformation, yielding, fracture and fatigue behavior of conventional and highly cross-linked ultra high molecular weight polyethylene. *Biomaterials* **2005**, *26*, 905–915.
39. Oral, E.; Malhi, A.S.; Muratoglu, O.K. Mechanisms of decrease in fatigue crack propagation resistance in irradiated and melted UHMWPE. *Biomaterials* **2006**, *27*, 917–925.
40. Muratoglu, O.K. Highly crosslinked and melted UHMWPE. In *UHMWPE Biomaterials Handbook*, 2nd ed.; Kurtz, S.M., Ed.; Academic Press: Burlington, MA, USA, 2009; pp. 197–204.

Chapter 11

Abrasion Resistance of Polymer Nanocomposites – A Review

Giulio Malucelli and Francesco Marino

Politecnico di Torino, DISMIC, Italy

INTRODUCTION

In order to be suitable for tribological applications, polymeric materials, which can usually exhibit mechanical strength, lightness, ease of processing, versatility and low cost, together with acceptable thermal and environmental resistances, have to show good abrasion and wear resistance. This target is not easy to achieve, since the viscoelasticity of polymeric materials makes the analysis of the tribological features and the processes involved in such phenomena quite complicated.

Indeed, it is well-known that an improvement of the mechanical properties can be effectively achieved by including "small" inorganic particles in the polymer matrices (Dasari et al., 2009).

For applications taking place in hard working conditions, such as slide bearings, the development of composite materials, which possess a high stiffness, toughness and wear resistance, becomes crucial. On the one hand, the extent of the reinforcing effect depends on the properties of composite components, and on the other hand it is strongly affected by the microstructure represented by the filler size, shape, homogeneity of distribution/dispersion of the particles within the polymer, and filler/matrix interface extension. This latter plays a critical role, since the composite material derives from a combination of properties, which cannot be achieved by either the components alone.

Thus it is generally expected that the characteristics of a polymer, added of a certain volume fraction of particles having a certain specific surface area, are more strongly influenced when very small particles (*nanofillers*), promoting an increased interface within the bulk polymer, are used (Bahadur, 2000; Chen et al., 2003; Karger-Kocsis & Zhang, 2005; Li et al., 2001; Sawyer et al., 2003). However, this happens only when a high dispersion efficiency of the nanoparticles within the polymer matrix is assessed: indeed, nanoparticles usually tend to agglomerate because of their high specific surface area, due the adhesive interactions derived from the surface energy of the material. In particular, the smaller the size of the nanoparticles, the more difficult the breaking down of such agglomerates appears, so that their homogeneous distribution within the polymer matrix is compromised.

As a consequence, the development of nanocomposites showing high tribological features requires a deep investigation on their micro-to-nanostructure, aiming to find synergistic mechanisms and reinforcement effects exerted by the nanofillers (Burris et al., 2007).

In addition, the way in which nanofillers can improve the tribological properties of polymers depends on the requirement profile of the particular application, i.e. the friction coefficient and the wear resistance cannot be considered as real material properties, since they depend on the systems in which these materials have to function.

In particular, such applications as brake pads or clutches usually require a high friction coefficient and, at the same time, a low wear resistance; however, in other circumstances (like in the case of gears or bearings, acting as smooth metallic counterparts under dry sliding conditions) the development of polymer composites having low friction and wear properties is strongly needed.

The abrasion performances of polymeric materials depend on several factors, such as the wear mechanisms involved, the abrasive test method used, the bulk and surface properties of the tested specimens, ...

Many papers reported in the literature focus on the investigation on the physical processes involved in abrasive wear of a wide variety of polymers; the obtained results demonstrate that two very different mechanisms of wear may occur in polymers, namely cohesive and interfacial wear processes, as schematically shown in Figure 1.

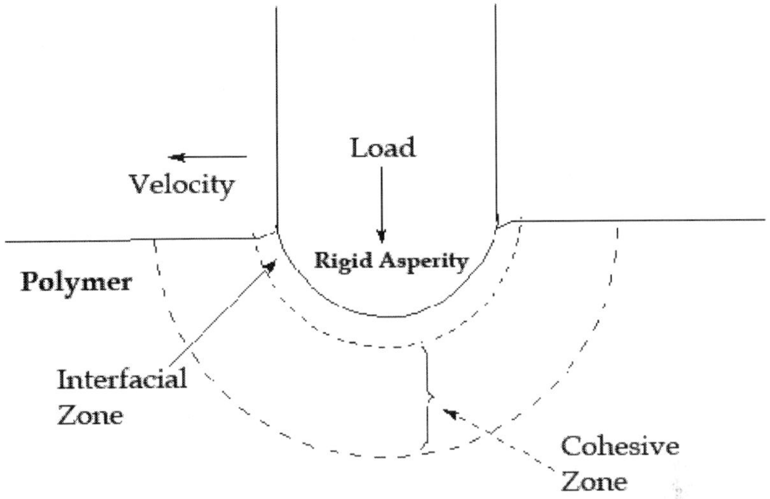

Figure. 1. Schematic representation of cohesive and interfacial wear processes (Adapted from Briscoe & Sinha, 2002)

In the cohesive wear processes, such as abrasion wear, fatigue wear and fretting, which mainly depend on the mechanical properties of the interacting materials, the frictional work involves quite large volumes close to the interface, either exploiting the interaction of surface forces and the consequent traction stresses or through the geometric interlocking exerted by the interpenetrating contacts. Contact stresses and contact geometry represent two key parameters that determine the extent of such surface zone.

On the other hand, the frictional work in interfacial wear processes (like transfer wear, chemical or corrosive wear) is dissipated in much thinner zones and at greater energy density with respect to cohesive wear processes, so that a significant increase in local temperature occurs. Furthermore, the extent of wear damage can be substantially ascribed to the chemistry of the surfaces involved, rather than to the mechanical properties of the interacting materials.

As far as cohesive processes are concerned, *abrasion wear*, which is the most common type of wear encountered in polymer composites, can be divided into two-body and three-body abrasion wear. The former occurs in the presence of hard asperities that plough and induce plastic deformation or fracture of the softer asperities.

The latter relates to the presence of hard abrasive particles or wear debris in between the sliding bodies: such particles or debris derive from environmental contaminants or can be the consequence of two-body abrasion processes. In general abrasion wear depends on several factors, such as the hardness of the materials in contact, the applied load and sliding distance and the geometry of the abrasive particles as well.

Fatigue wear derives from surface fatigue phenomena, i.e. from the repeated stressing and un-stressing of the contacts, and can lead to fracture through the accumulation of irreversible changes, which determine the generation, growth and propagation of cracks. This kind of wear may also occur together with *delamination wear*, where shear deformations of the softer surface, caused by traction of the harder asperities, promote the nucleation and coalescence of subsurface cracks. As a consequence, the delamination (i.e. detachment) of fragments having larger size occurs.

Fretting wear is caused by relative oscillatory motions of small amplitude taking place between two surfaces in contact. The produced wear fragments can either escape from between the surfaces, thus promoting a fit loss between the surfaces and a decrease of clamping pressure, which may lead to higher vibration effects, or remain within the sliding surfaces, so that pressure increases and seizure eventually occurs.

Transfer wear belongs to interfacial wear processes and involves the formation of a transfer film (solid or liquid, depending on the interfacial temperature) in polymer-metal, polymer- ceramic, polymer-polymer sliding contacts. Such film invariably transfers from polymer to metal or ceramic, whereas the direction of transfer is not obvious in the case of polymer-polymer sliding contacts.

Several parameters can influence the formation of the transfer film and its role on the subsequent wear processes: thickness and stability of the film, cohesion features between the transfer layers, adhesion forces between the film and the sliding counterpart, chemical reactivity and surface roughness of the counterface slider, polymer structure (crystallinity, flexibility, presence of pendant groups or side chains, …), adopted sliding conditions (temperature, normal load, velocity, atmosphere, …) and presence of fillers.

Chemical wear involves a chemical reaction (oxidation, degradation, hydrolysis, …, which lead to polymer chain scission with the subsequent MW decrease) in between the sliding bodies or a material in itself or a material with the surrounding environment.

A schematic representation of the basic tribological interactions leading to wear particle generation is depicted in Figure 2.

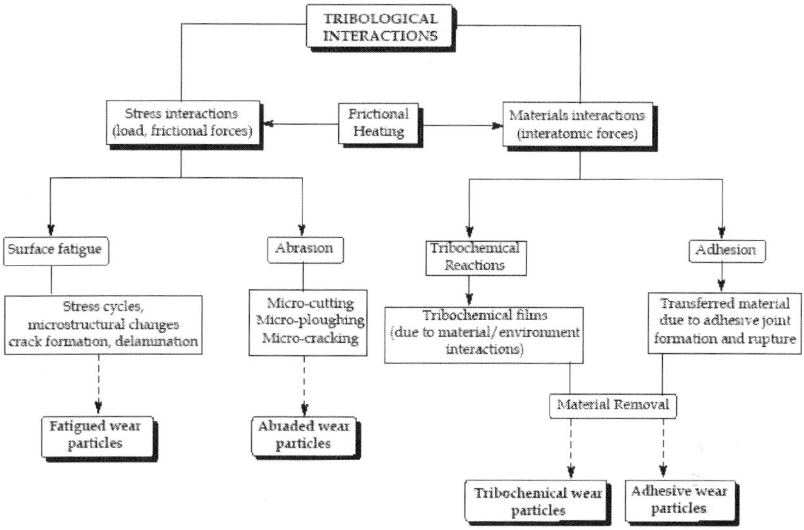

Figure. 2. Different wear processes leading to the formation of material particles (adapted from Czichos, 2001)

It is worthy to note that the wear mechanisms in polymer systems described above for macro- and micro-levels are quite different from those encountered at nano-level.

First of all, nano-level involves very low applied loads (from □N to nN); in addition, the wear particle generation is negligible and the original surface topography is more likely to be preserved for an extended period because of the adopted low wear rate.

Other differences concern the friction forces involved at the nano-level, since the ploughing factor and the inertial effect of the moving components are different, as well as the role exerted by surface forces (adhesion and electrostatic interactions), which become very important.

In the following paragraphs, a review on the recent studies on the tribological behavior of thermoplastic nanocomposites is presented. The role of the structure of the nanofillers and of their morphology (aspect ratio, effectiveness of dispersion within the polymer, …) and the possible interactions with the environment are widely discussed.

TRIBOLOGY OF THERMOPLASTIC NANOCOMPOSITES

PEEK-based nanocomposites

Poly(ether ether ketone) (PEEK) is a high performance injection mouldable thermoplastic that can be widely used for many applications that require high mechanical strength and an outstanding thermo-mechanical stability. This polymer has a high glass transition temperature (Tg≈143°C) and a high melting point (Tg≈343°C) and it is also regarded as one of the most promising polymer materials for tribological applications in aqueous environments.

Nevertheless, it seems that neat PEEK exhibits relatively poor wear resistance with water lubrication in some cases, so that different types of fillers (and nanofillers) have been incorporated into this polymer, aiming to facilitate more applications by enhancing its anti- wear features. In particular, short carbon fibers (SCFs) are currently used in PEEK-based composites for improving its wear resistance, even at elevated temperatures and under aqueous conditions (water lubrication).

Very recently, Zhong investigated the tribological properties of PEEK/SCF/zirconia composites under aqueous conditions, using a three-pin-on-disc configuration (Zhong et al., 2011). A synergistic effect of SCFs with zirconia nanoparticles was assessed: indeed, the composites showed excellent wear resistance under aqueous conditions; SCFs were found to carry the main load between the contact surfaces and to protect the polymer matrix from further severe abrasion of the counterpart. Nano-ZrO2 efficiently inhibited SCF failure either by reducing the stress concentration on the CF interface through reinforcement of the matrix or by lowering the shear stress between the sliding surfaces via a positive rolling effect of the nanoparticles between the material pairs.

Werner et al. investigated the influence of vapour-grown carbon nanofibres (CNFs) on the wear behaviour of PEEK (Werner et al., 2004). To this aim, unidirectional sliding tests against two different counterpart materials (100Cr6 martensitic bearing steel and X5CrNi18-

10 austenitic stainless steel) were performed on injection moulded PEEK-CNF nanocomposites. CNFs were found to reduce the wear rate of PEEK very significantly, as compared to a variety of commercial PEEK grades. This behaviour was attributed to CNFs, which act as solid lubricants; in addition, the roughening effect on the counterpart exerted by CNFs, because of their small size, was minimised with respect to conventional fibre fillers (carbon fibres, PAN-based carbon fibres, glass fibres).

McCook and coworkers investigated the role of different micro and nanofillers on the tribological properties of PEEK in dry sliding tests against 440C stainless steel counterfaces (McCook et al., 2007). To this aim, microcrystalline graphite, carbon nano-onions, single- walled carbon nanotubes, C60 carbon fullerenes, microcrystalline WS2, WS2 fullerenes, alumina nanoparticles and PTFE nanoparticles were jet-milled with PEEK and the friction coefficients and wear rates of the obtained composites were measured in open laboratory air (45% R.H.) and in a dry nitrogen environment (less than 0.5% R.H.).

Both wear rate and friction coefficient were reduced in the dry nitrogen environment: in particular, the more wear resistant coatings also had lower friction coefficients. On the contrary, in open air environments the more wear resistant coating exhibited the higher friction coefficients. Furthermore, the polymeric nanocomposites investigated showed similar environmental responses, regardless of the type of micro or nanofillers used.

Hou and coworkers performed tribological ball-on-flat sliding wear tests on PEEK-based nanocomposites incorporating inorganic fullerene-like tungsten disulfide nanoparticles (Hou et al., 2008). The friction coefficient was found to decrease about 3 times in the presence of 2.5 wt.% nanoparticles, with respect to the neat PEEK: this behaviour was attributed to the lubricating capability of the nanofillers.

Zhang et al. investigated the effect of nano-silica particles on the tribological behaviour of PEEK: silica nanoparticles were compounded with the polymer by means of a ball milling technique (Zhang et al., 2008). The wear resistance of PEEK was significantly improved after incorporating nano-SiO2 and at a rather low filler loading (1 vol.%), the composites showed the optimum wear resistance, which was ascribed to the reduced perpendicular deformation of PEEK matrix and to the decreased tangential plastic flow of the surface layer involved in friction processes. Furthermore, the nanocomposites evidenced much smoother surfaces with respect to neat PEEK.

Pursuing this research, the role of the same nano-silica particles on the tribological behaviour of SCF-reinforced PEEK was also investigated (Zhang et al., 2009). To this aim, 1 vol.% (1.51 wt.%) nano-SiO2 particles were compounded with SCF/PTFE/graphite filled PEEK in a Brabender mixer; the obtained composite materials were tested using a block-on- ring apparatus at room temperature (counterpart: 100Cr6 steel ring), in extremely wide pressure and sliding velocity ranges. Under all the conditions investigated, nano-SiO2 particles remarkably reduced the friction coefficients; above 2 MPa pressures, the nanoparticles were found to reduce the wear rate: this behaviour was attributed to a protection effect of SCF/PEEK interface exerted by the nanoparticles, which are able to reduce the stress concentration on SCFs taking place in the surface layer involved into friction.

Zhang also investigated the effect of different amounts of nano-silica particles on the tribological behavior of SCF-reinforced PEEK composites. The nanoparticle loading was varied from 1 to 4 vol.% (Zhang et al., 2009).

The variation of nanoparticle content from 1 to 4 vol.% did not significantly affect the friction coefficients of the nanocomposites; in addition, operating with low pressure-sliding velocity (pv) factors, the nanoparticles turned out to worsen the wear rate of the composites, because of the abrasion on SCFs exerted by nanoparticle agglomerates. On the contrary, with a high pv factor, such agglomerates were crushed into tiny ones, so that nano-silica particles were capable to protect SCFs reducing their failures. Similar wear rates were found for the nanocomposites tested at very high pv factors.

Polyolefin-based nanocomposites

Thermoplastic polyolefins like poly(ethylene)s (PEs) and poly(propylene) (PP) are well- established polymers available at the market, each having a different structure and very different behaviour, performances and applications (Feldman & Barbalata, 1996). Several papers deal with their tribological properties, in the presence of different types of nanofillers.

High density poly(ethylene) (HDPE) was used as matrix for preparing nanosilica coatings, the wear resistance of which was measured using a rotative drum abrader (Barus et al., 2009). It was found that this parameter, despite a significant increase in the mechanical properties of the nanocomposites (stiffness, yield strength and fracture toughness), exhibited lower values with respect to the neat polymer.

Johnson and coworkers manufactured and tested the wear behaviour of HDPE/multi- walled-carbon-nanotubes composites (Johnson et al., 2009). Different weight percentages of nanotubes (1, 3 and 5%) were used for preparing the samples, which were tested on a block- on-ring apparatus. Wear resistance and frictional properties of HDPE were found to improve in the presence of the nanofillers; furthermore, the addition of multi-walled- carbon-nanotubes to HDPE turned out to bring wear rates down to the level seen in ultra- high molecular weight poly(ethylene) (UHMWPE).

The effect of the presence of Alumina nanoparticles (5 wt.%) was exploited for investigating the abrasion resistance of low-density poly(ethylene) (LDPE)-based nanocomposites (Malucelli et al., 2010). The abrasion resistance of the nanocomposites increased in the presence of the nanofillers, as indicated by the decrease of the Taber Wear Index with respect to the neat polymer.

Very recently, Xiong and coworkers investigated the effect of the presence of nano- hydroxyapatite (nano-HAP) on the tribological properties of non-irradiated and irradiated UHMWPE composites, prepared by using a vacuum hot-pressing method (Xiong et al., 2011). The friction coefficients and wear rates were measured by using a reciprocating tribometer (counterface: CoCr alloy plates). The presence of 7 wt.% nano-HAP in the polymer matrix resulted in lowering both the friction coefficients and wear rate, irrespective of using irradiated or non-irradiated samples, whereas filling 1 wt.% nano-HAP reduced friction coefficients and wear rate of the non-irradiated UHMWPE only.

Misra and coworkers investigated the tribological behaviour of polyhedral oligomeric silsesquioxanes (POSS)/poly(propylene) nanocomposites (Misra et al., 2007). The relative friction coefficient of the samples turned out to strongly decrease from 0.17 for neat PP to

0.07 for the nanocomposite containing 10 wt.% POSS: this behaviour was ascribed to the increase of the surface hardness and of the modulus, due to the presence of the nanofiller.

Fluorinated-based nanocomposites

Fluorinated polymers usually exhibit many desirable tribological features, including low friction, high melting temperature and chemical inertness. However, their anti-wear applications have been somewhat limited by their poor wear resistance, which has led to the failure of anti-wear components and films.

Therefore, many researchers have tried to reinforce fluorinated polymers using different fillers, such as glass fibres, carbon fibres, ceramic powders, non-ferrous metallic powders: unfortunately, these fillers induced a large frictional coefficient and abrasion. Quite recently, nanometer size inorganic powders have been chosen as fillers capable to enhance the wear behaviour of fluorinated polymers.

Poly(tetrafluoroethylene), PTFE, is the most common fluorinated polymer used for tribological purposes.

Lee and coworkers added carbon-based nanoparticles, synthesized by heat treatment of nanodiamonds, to PTFE, in order to prepare fluorinated nanocomposites (Lee et al., 2007). The wear resistance, measured through ball-on-plate wear tests, was found to depend on the heat treatment, which nanodiamonds were subjected to: in particular, wear resistance turned out to increase when nanodiamonds were heated at 1000°C. Beyond this temperature, carbon nanoparticles became aggregated and therefore the wear coefficient of the obtained nanocomposites increased: this failure in the wear behaviour was ascribed to the formation of carbon onions that promoted the aggregation of carbon nanoparticles.

Single-walled carbon nanotubes have been exploited for lowering the wear rates of PTFE (Vail et al., 2009). A linear reciprocating tribometer was exploited for performing the tests (counterface: 304 stainless steel) on nanocomposite samples containing up to 15 wt.% nanotubes. The obtained results clearly indicated that, in the presence of low nanofiller loadings (5 wt.%), PTFE wear resistance is improved by more than 2000% and friction coefficient increased by ≈50%.

Shi and coworkers have studied the effect of various filler loadings (from 0.1 to 3 wt.%) on the tribological properties of carbon-nanofiber (CNF)-filled PTFE composites (Shi et al., 2007). The friction and wear tests were conducted on a ring-on-ring friction and wear tester. The counterface materials was steel 45.

The obtained results showed that the friction coefficients of the PTFE composites decreased initially up to a 0.5 wt.% filler concentration (during sliding, the released CNFs transfer from the composite to the interface between the mating surfaces, acting as spacers and thus preventing direct contact between the two surfaces and lowering the friction coefficient) and then increased, whereas the anti-wear properties of the materials increased by 1-2 orders of magnitude in comparison with those of PTFE. Finally, the composite having 2 wt.% of CNFs exhibited the best anti-wear properties under all the experimental friction conditions.

The tribological investigation on fluorinated polymers has been also extended to PTFE- based blends, as described by Wang and coworkers (Wang et al., 2006). In particular, Xylan 1810/D1864, a commercially available PTFE blend for dry lubricant and corrosion resistant coatings, has been blended with alumina nanoparticles at different loadings (from 5 to 20 wt.%). The wear resistance was measured using a Taber Abrasion Tester and was found to decrease with increasing the content of the embedded alumina nanoparticles in the polymer matrix. The minimum wear rate was achieved when the nanoparticle loading was 20 wt.%.

Another paper from Burris and Sawyer reports on the role of irregular shaped alumina nanoparticles on the wear resistance of Al2O3/PTFE nanocomposites (Burris & Sawyer, 2006). A reciprocating pin-on-disc tribometer was used for testing the wear and friction of the samples (counterface: AISI 304 stainless steel plates). It was found that the inclusion of irregular shaped alumina particles is more effective in reducing PTFE wear than spherical shaped particles (the wear resistance of PTFE was increased 3000x in the presence of 1 wt.% former nanofiller), but also determines an increased friction coefficient.

Another fluorinated polymer, namely poly(vinylidene fluoride), was used as matrix for preparing nanocomposites containing a phyllosilicate (organoclay) by Peng and coworkers (Peng et al., 2009). The friction and wear tests were conducted on different loaded nanocomposites (clay content: 1 - 5 wt.%), using a block-on-ring wear tester (mated ring specimen: carbon steel 45, GB 699-88). The nanoclay at 1-2 wt.% turned out to be effective for improving the tribological properties of neat PVDF, since such filler may act as a reinforcement to bore load and thus decrease the plastic deformation.

Tribological studies were also performed on PTFE-based fabric composites (Sun et al., 2008; Zhang et al., 2009). In particular, Sun and coworkers prepared polyester fabric composites, in order to study the influence of alumina nanoparticles and PTFE micro-powders embedded in an epoxy matrix on the tribological properties of the fabric composites. The excellent tribological performance of the fillers significantly turned out to enhance the wear resistance of the fabric polyester composites.

Zhang et coworkers were able to improve the wear resistance of PTFE/phenolic/cotton fabric composites, by dispersing functionalized multi-walled carbon nanotubes in the phenolic resin (Zhang et al., 2009). Sliding tests were performed on a pin-on-disc tribometer (flat-ended AISI-1045 pin). The high homogeneity of dispersion of the nanofiller allowed to achieve an improved wear resistance in the fabric composites; furthermore, the tribological properties of the obtained systems were found to strongly depend on the carbon nanotubes content: 1 wt.% nanofiller was the optimum loading for maximizing the wear resistance of the fabric composites.

Poly(amide)-based nanocomposites

Poly(amide) 6 and 66 (Nylon 6 and Nylon 66) have been widely used as engineering plastics in different applications, such as bearings, gears or packaging materials. They possess an outstanding combination of properties such as high toughness, tensile strength and abrasion resistance, low density and friction coefficient and quite easy processing. Indeed, their abrasion resistance is a key factor for their widespread applications.

Aiming to further improve their mechanical properties and tribological behaviour, nylons were reinforced with some micro-particles or fibres, such as CuS, CuF2, CuO, PbS, CaO, CaS and carbon fibres: they were effective in reducing the wear rate of polyamides (Bahadur et al., 1996).

In quite recent years, as for other thermoplastic matrices, several nano-materials were served as suitable fillers of poly(amides) for improving their integrated properties, particularly referring to their tribological behavior.

Garcia and coworkers found that nano-SiO2 could reduce effectively the coefficient of friction and wear rate of nylon 6: in particular, the addition of 2 wt.% nano-SiO2 determined the lowering of the friction coefficient from 0.5 to 0.18 (Garcia et al., 2004). This was possible since the surface of nylon 6 nanocomposites was well protected by the transfer film on the surface of the metal counterface. At the same time, the low silica loading led to a reduction in wear rate by a factor of 140, whereas the effect of higher silica loadings was less pronounced.

Dasari and coworkers reported on the role of nanoclays on the wear characteristics of nylon 6 nanocomposites processed via different routes (Dasari et al., 2005). They demonstrated that aggregated nanoclay particles result in the worst wear resistance of the nanocomposites, whereas the systems, which exhibit a good interfacial adhesion of clay to polymer matrix, together with an homogeneous clay dispersion, determine substantial improvements of wear resistance.

Zhou and coworkers investigated the tribological behaviour of Nylon 6/Montmorillonite clay nanocomposites: the poor abrasion resistance exhibited by the nanocomposites was attributed to the presence of defects at the clay/polymer interface, resulting in lower wear resistance of the polymer matrix as the nanofiller content increased (Zhou et al., 2009).

Sirong and coworkers studied the tribological behaviour of Nylon 66/organoclay nanocomposites, in the presence of styrene-ethylene/butylene-styrene triblock copolymer grafted with maleic anhydride (SEBS-g-MA) as a toughening agent (Sirong et al., 2007). A pin-on-disc friction and wear testing apparatus was used in sliding experiments (counterface: 65 HRC steel disc). It was demonstrated that the use of SEBS-g-MA allows to obtain significant improvements as far as the wear resistance of the nanocomposite is concerned: this behaviour was ascribed to the toughening effect of SEBS-g-MA, which favours the transfer of a uniform, continuous and smooth thin film to the steel counterface, thus avoiding the direct contact of this latter with the nanocomposite.

Poly(amide) 66 was also chosen as matrix for preparing nanoparticle-filled composites (Chang et al., 2006). Different fillers, such as TiO2 nanoparticles (5 vol.%), short carbon fibres (15 vol.%) and graphite flakes (5 vol.%), were added to the polymer and the obtained composites tested on a pin-on-disc apparatus (counterface: polished steel disc). It was found that nano-TiO2 could effectively reduce the frictional coefficient and wear rate, especially under higher pv conditions. In order to further understand the wear mechanisms, the worn surfaces were examined by scanning electron microscopy and atomic force microscopy; a positive rolling effect of the nanoparticles between the material pairs was proposed, which contributes to the remarkable improvement of the load carrying capacity of polymer nanocomposites.

Quite recently, Ravi Kumar and coworkers studied the synergistic effect of nanoclay and short carbon fibers on the abrasive wear behavior of nylon 66/poly(propylene) nanocomposites (Ravi Kumar et al., 2009). A modified dry sand rubber wheel abrasion tester was employed for performing the three-body abrasive wear experiments. The obtained results clearly indicated that the addition of nanoclay/short carbon fiber in PA66/PP significantly influences wear under varied abrading distance/loads. Furthermore, it was found that nanoclay filled PA66/PP composites exhibited lower wear rates with respect to short carbon fiber filled PA66/PP composites.

Poly(oxymethylene)-based nanocomposites

Poly(oxymethylene) (POM) is an engineering polymer that has been widely used as self- lubricating material for many applications, such as automobile, electronic appliance and engineering. This polymer exhibits good fatigue resistance, creep resistance and high impact strength. Its low friction coefficient derives from the flexibility of the linear macromolecular chains; in addition, its high crystallinity and high bond energy result in good wear resistant properties. Some papers report on the preparation of polymeric nanocomposites based on POM.

Various fillers or fibers, such as graphite, MoS2, Al2O3, PTFE, glass and carbon fibers, have been incorporated into POM matrices as internal lubricants or reinforcements to further enhance the tribological properties of such a polymer.

Kurokawa et al. investigated the tribological properties of POM composites containing very small amounts of silicon carbide (SiC) and/or calcium salt of octacosanoic acid (Ca-OCA), as well as PTFE (Kurokawa et al., 2000). It was found that the incorporation of Ca-OCA into POM/SiC composites drastically lowered their friction coefficient; furthermore, the wear rate was also lowered because of the nucleating effect of SiC and Ca-OCA.

Wang and coworkers prepared POM/MoS2 nanocomposites by in situ intercalation polymerization: the intercalated composites showed a significant decrease of friction coefficient, together with an improved wear resistance, especially under high load, while the heat resistance of the composites decreased slightly (Wang et al., 2008).

The same research group also prepared POM/ZrO2 nanocomposites, which evidenced better wear resistance with respect to neat POM, whereas the change in friction coefficient of the nanocomposites was very limited. (Wang et al., 2007)

Sun and coworkers studied the tribological properties of POM/Al2O3 nanocomposites (Sun et al., 2008). The friction and wear measurements were conducted on a friction and wear tester, using a block-on-ring arrangement (counterface: HRC50-55 plain carbon steel ring). It was found that alumina nanoparticles were more effective in enhancing the tribological properties of Poly(oxymethylene) nanocomposites in oil lubricated condition rather than in dry sliding experiments. Indeed, the former environment allows to form a uniform and compact transfer film on the surface of the counterpart steel ring, whereas the transfer film under dry sliding condition is destroyed by the agglomerated abrasives residing between the friction surfaces. The optimal nanoparticles content in POM nanocomposites was 9% under oil lubricated condition, below which alumina nanoparticles between the friction surfaces were still under saturation.

Sun and coworkers have also investigated the tribological behaviour of Poly(oxymethylene) (POM) composites compounded with nanoparticles, PTFE and MoS2 in a twin-screw extruder (Sun et al., 2008). The tribological tests were performed on a friction and wear tester using a block-on-ring arrangement under dry sliding and oil lubricated conditions, respectively. The better stiffness and tribological properties exhibited by POM nanocomposites with respect to POM composites were attributed to the high surface energy of the nanoparticles; the only exception was represented by the decreased dry-sliding tribological properties of POM/3%Al2O3 nanocomposite, ascribed to Al2O3 agglomeration. Furthermore, the friction coefficient and wear volume of POM nanocomposites under oil lubricated condition decreased significantly.

Poly(methylmethacrylate)-based nanocomposites

Poly(methylmethacrylate), PMMA, is an important engineering polymer, which finds application in many sectors such as aircraft glazing, signs, lighting, architecture, and transportation. In addition, since PMMA is non-toxic, it could be also useful in dentures, medicine dispensers, food handling equipment, throat lamps, and lenses.

Unfortunately, this polymer shows poor abrasion resistance with respect to glass, thus limiting its potential use in other fields. Despite several efforts, attempts to improve the PMMA scratch and abrasion resistance have induced other drawbacks, such as a decrease of the impact strength, so that researchers focused on the preparation of PMMA nanocomposites.

Avella and coworkers studied the tribological features of PMMA-based nanocomposites filled with calcium carbonate (CaCO3) nanoparticles, exploiting in situ polymerization (Avella et al., 2007). In order to improve inorganic nanofillers/polymer compatibility, poly(butylacrylate) chains have been grafted onto CaCO3 nanoparticle surface.

$CaCO_3$ nanoparticles, regardless of the presence of the grafting agent, turned out to significantly improve the abrasion resistance of PMMA also modifying its wear mechanism: indeed, the nanoparticles induced only micro-cutting and/or micro-ploughing phenomena, thus generating a plastic deformation and consequently increasing the abrasion resistance of the polymer matrix.

The same research group also investigated the tribology of PMMA-based nanocomposites containing modified silica nanoparticles, obtained through in situ polymerization approach (Avolio et al., 2010). The high compatibility between silica nanoparticles and the polymer allowed to significantly improve the abrasion resistance of PMMA, because nanoparticles were able to support part of the applied load, thus reducing the penetration of grains of the rough abrasive wheel into PMMA surface and contributing to the wear resistance of the material.

Dong and coworkers prepared Poly(methyl methacrylate)/styrene/multi-walled carbon nanotubes (PMMA/PS/MWNTs) copolymer nanocomposites by means of in situ polymerization method (Dong et al., 2008). The tribological behavior of the copolymer nanocomposites was investigated using a friction and wear tester under dry conditions: with respect to pure PMMA/PS copolymer, the copolymer nanocomposites showed not only better wear resistance but also smaller friction coefficient. MWNTs were found to strongly improve the wear resistance property of the copolymer nanocomposites, because of their self-lubricating features, their homogeneous and uniform distribution within the copolymer matrix and their help in forming thin running MWNTs films that slide against the transfer film (developed on the surface of the stainless steel counterface).

Very recently, Carrion and coworkers exploited single-walled carbon nanotubes modified with an imidazolium ionic liquid for preparing PMMA nanocomposites and studying their dry tribological performances as compared to neat PMMA or to the nanocomposites containing pristine carbon nanotubes without ionic liquid (Carrion et al., 2010). The tribological behavior of the obtained nanocomposites, studied against AISI 316L stainless steel pins, resulted in a significant wear rate decrease with respect to PMMA/carbon nanotubes (-58%) and neat PMMA (-63%).

Other thermoplastic-based nanocomposites

Some other thermoplastic engineering and specialty polymers have been considered as far as tribological issues are concerned. In the following, we will summarize the recent progress in understanding wear and friction in nanocomposite systems based on these polymers.

Bhimaraj and coworkers studied the friction and wear properties of poly(ethylene) terephthalate (PET) filled with alumina nanoparticles (up to 10 wt.% nanofiller), using a reciprocating tribometer (Bhimaraj et al., 2005). The obtained results showed that the addition of alumina nanoparticles can increase the wear resistance by nearly 2x over the unfilled polymer. Furthermore, the average friction coefficient also decreased in many cases. This behavior was attributed to the formation a more adherent transfer film that protects the sample from the steel counterface, although the presence of an optimum filler content could be ascribed to the development of abrasive agglomerates within the transfer films in the higher wt.% samples.

Another paper from the same research group reports on the effect of particle size, loading and crystallinity on PET/Al2O3 nanocomposites (Bhimaraj et al., 2008). The nanocomposite samples were tested in dry sliding against a steel counterface. The tribological properties were found to depend on crystallinity, filler size and loading; in addition, wear rate and friction coefficient were very low at optimal loadings that ranged from 0.1 to 10 wt.%, depending on the crystallinity and particle size.

Wear rate were found to lower monotonically with decreasing particle size and crystallinity at any loading in the range tested.

Poly(etherimide)s (PEIs) are high-performance thermoplastics with high modulus and strength, superior high temperature stability, as well as electrical (insulating) and dielectric properties (very low dielectric constant). These polymers perform successfully in aerospace, electronics, and other applications under extreme conditions. Nevertheless, pure PEIs show such disadvantages as brittleness and high wear rate, which limit their applications. Therefore, appropriate modifications of PEIs with nanofillers have been proposed, in order to widen their industrial applications.

Chang and coworkers reinforced PEI with titania nanoparticles, in the presence of short carbon fibres (SCFs) and graphite flakes as well (Chang et al., 2005). Wear tests were performed on a pin-on-disc apparatus, using composite pins against polished steel counterparts, under dry sliding conditions, different contact pressures and various sliding velocities. SCFs and graphite flakes turned out to remarkably improve both the wear resistance and the load-carrying capacity. Nevertheless, the addition of nano-TiO2 further reduced the frictional coefficient and the contact temperature of the composites, especially under high pv conditions.

The same research group investigated the role of the presence of nano- or micro-sized inorganic particles (5 vol.% nano TiO2 or micro-CaSiO3) on the tribological behavior of PEI matrix composites, additionally filled with SCFs and graphite flakes (Xian et al., 2006). The influence of these inorganic particles on the sliding behavior was assessed with a pin-on-disc tester at room temperature and 150°C.

The obtained results showed that both micro and nano particles could reduce the wear rate and the friction coefficient of the PEI composites under the experimental adopted conditions, but in a different temperature range: indeed, the microparticles filled composites showed improved tribological features at room temperature, whereas the nano-titania-filled composites possessed the lowest wear rate and friction coefficient at elevated temperature. The tribological improvements evidenced by the nano-particles were attributed to the formation of transfer layers on both sliding surfaces together with the reinforcing effect.

Very recently, Li and coworkers dispersed carbon nanofibers (from 0.5 to 3 wt.%) in a PEI matrix through a melt mixing method and tested the tribological properties of the obtained nanocomposites (Lee et al., 2010). The composites containing 1 wt.% CNFs showed very high wear rates comparable with that of pure PEI; nevertheless, higher CNF loadings promoted a significant reduction in wear rate at steady state wear.

Like PMMA, also poly(carbonate) (PC), an amorphous engineering thermoplastic, which combines thermal stability, good optical properties, outstanding impact resistance and easy processability, shows poor scratch and abrasion resistance with respect to glass, thus limiting its potential use in fields other than medical, optics, automotive, ….

Carrion and coworkers prepared a new polycarbonate nanocomposite containing 3 wt.% organically modified nanoclay by extrusion and injection moulding, and its tribological properties were measured under a pin-on-disc configuration against stainless steel (Carrion et al., 2008). The obtained nanocomposites showed 88% of reduction in friction coefficient and up to 2 orders of magnitude reduction in wear rate with respect to the neat polymer. Such good tribological performances were attributed to the uniform microstructure achieved and to the nanoclay intercalation.

CONCLUSION

The significant spreading of research activities concerning the tribology of thermoplastics and thermoplastic-based nanocomposites demonstrates that this topic is very up-to-date. Indeed, several low-loading, low-wear polymer nanocomposites are being prepared and evaluated in tribology laboratories.

In many cases, nanocomposite systems result in outperforming traditional macro- and micro-composites by orders of magnitude with substantially lower filler loadings (often less than 5 wt.%), provided that the tribological features strongly depend on the homogeneity of dispersion and distribution of the nanofillers within the polymer matrix.

Past macro and micro models, which have been always exploited for estimating the mechanical behavior of composite materials seem to be quite inadequate to describe the phenomena occurring at a nanoscale level, particularly referring to wear and friction.

The standard tools applied for characterizing nanomaterials need to be implemented more in tribology studies to help clarify the obtained experimental results. This means that tribology should always be considered as an important issue of the materials science.

In particular, regardless of the effectiveness of the nanofiller dispersion within the polymer matrix, some issues become very crucial and should be consequently deeply investigated. First of all, the chemistry and chemical reactions, which may occur in between the mating surfaces, have to be considered, and the influence of the by-products resulting from such reactions or during wear as well.

Indeed, the effect and dynamics of the development of the transfer film during low wear sliding, together with the evolution of its physico-chemical and mechanical properties should be thoroughly investigated. Consequently, the mechanisms, through which removal of abraded materials occurs, should be deeply investigated, so that proper mechanics models for the design of high wear resistant nanocomposites can be developed.

Finally, synergies between materials science and tribology have to be developed, aiming to better understand the complex tribological phenomena taking place in polymeric nanocomposites.

This approach will surely contribute to design more efficient nanomaterials for tribological applications.

ACKNOWLEDGMENT

The financial support of Piedmont Region, Italy (Innovative Systems for Environmental friendly air COMPression – ISECOMP Project 424/09 – Piedmont Region industrial research call 2008) is gratefully acknowledged.

REFERENCES

1. Avella, M.; Errico, M.E., Gentile, G. (2007). PMMA Based Nanocomposites Filled with Modified CaCO3 Nanoparticles. *Macromolecular Symposia,* Vol.247, pp. 140-146
2. Avolio, R.; Gentile, G., Avella, M., Capitani, D., Errico, M.E. (2010). Synthesis and Characterization of Poly(methylmethacrylate)/Silica Nanocomposites: Study of the Interphase by Solid-State NMR and Structure/Properties Relationships. *Journal of Polymer Science: Part A: Polymer Chemistry,* Vol.48, pp. 5618-5629
3. Bahadur, S.; (2000). The Development of Transfer Layers and their Role in Polymer Tribology. *Wear,* Vol.245, pp. 92-99
4. Bahadur, S.; Gong, D., Anderegg, J. (1996). Investigation of the

Influence of CaS, CaO and CaF2 Fillers on the Transfer and Wear of Nylon by Microscopy and XPS Analysis. *Wear,* Vol.197, pp. 271-279

5. Barus, S.; Zanetti, M., Lazzari, M., Costa, L. (2009). Preparation of Polymer Hybrid Nanocomposites Based on PE and Nanosilica. *Polymer,* Vol.50, pp. 2595-2600
6. Bhimarai, P.; Burris, D.L., Action, J., Sawyer, W.G, Toney, C.G., Siegel, R.W., Schadler, L.S. (2005). Effect of Matrix Morphology on the Wear and Friction Behavior of Alumina Nanoparticle/poly(ethylene) terephthalate Composites. *Wear,* Vol.258, pp. 1437- 1443
7. Bhimarai, P.; Burris, D.L., Action, J., Sawyer, W.G, Toney, C.G., Siegel, R.W., Schadler, L.S. (2008). Tribological Investigation of the Effects of Particle Size, Loading and Crystallinity on Poly(ethylene) terephthalate Nanocomposites. *Wear,* Vol.264, pp. 632-637
8. Briscoe, B.J.; Sinha, S.K. (2002). Wear of Polymers. *Proceedings of the Institution of Mechanical Engineers, Part J: Journal of Engineering Tribology,* Vol.216, pp. 401-413
9. Burris, D.L.; Sawyer, W.G. (2006). Improved Wear Resistance in Alumina-PTFE Nanocomposites with Irregular Shaped Nanoparticles. *Wear,* Vol.260, pp. 915-918
10. Burris, D.L.; Boesl, B., Bourne, G.R., Sawyer, W.G. (2007). Polymeric Nanocomposites for Tribological Applications. *Macromolecular Materials and Engineering,* Vol.292, pp. 387-402
11. Carrion, F.J.; Arribas, A., Bermudez, A.K., Guillamon, A. (2008). Physical and tribological Properties of a New Polycarbonate-organoclay Nanocomposite. *European Polymer Journal ,* Vol.44, pp. 968-977
12. Carrion, F.J.; Espejo, C., Sanes, J., Bermudez, A.K. (2010). Single-walled Carbon Nanotubes Modified by Ionic Liquid as Antiwear Additives of Thermoplastics. *Composite Science and Tecnology,* Vol.70, pp. 2160-2167
13. Chang, L.; Zhang, Z., Zhang, H., Friedrich, K. (2005). Effect of Nanoparticles on the Tribological Behaviour of Short Carbon Fibre Reinforced poly(etherimide) Composites. *Tribology International,* Vol.38, pp. 966-973
14. Chang, L.; Zhang, Z., Zhang, H., Schlarb, A.K. (2006). On the Sliding Wear of Nanoparticle Filled Polyamide 66 Composites. *Composite Science and Technology,* Vol.66, pp. 3188- 3198
15. Chen, W.; Li, F., Han, G., Xia, J., Wang, L., Tu, J., Xu, Z. (2003). Tribological Behavior of Carbon-nanotube-filled PTFE Composites. *Tribology Letters*, Vol.15, pp. 275-278
16. Czichos, H. (2001). Tribology and Its Many Facets: From Macroscopic to Microscopic and Nano-scale Phenomena. *Meccanica,* Vol.50, pp. 605-615

17. Dasari, A.; Yu, Z.Z., Mai, Y.K. (2009). Fundamental Aspects and Recent Progress on Wear/scratch Damage in Polymer Nanocomposites. *Materials Science and Engineering R,* Vol.63, pp. 31-80
18. Dasari, A.; Yu, Z.Z., Mai, Y.K., Hu G.H., Varlet, J. (2005). Clay Exfoliation and Organic Modification on Wear of Nylon 6 Nanocomposites Processed by Different Routes. *Composite Science and Technology,* Vol.65, pp. 2314-2328
19. Dong, B.; Wang, C., He, B.L., Li, H.L. (2008). Preparation and Tribological Properties of Poly(methyl methacrylate)/Styrene/MWNTs Copolymer Nanocomposites. *Journal of Applied Polymer Science,* Vol.108, pp. 1675-1679
20. Feldman, D. & Barbalata, A. (1996). *Synthetic Polymers : Technology, Properties, Applications,* Springer, New York, USA
21. Garcia, M.; de Rooij, M., Winnbust, L., van Zyl, W.E., Verweij, H. (2004). Friction and Wear Studies on Nylon 6/SiO2 Nanocomposites. *Journal of Applied Polymer Science,* Vol.92, pp. 1855-1862
22. Hou, X.; Shan, C.X., Choy, K.W. (2008). Microstructures and Tribological Properties of PEEK-based Nanocomposites Coatings Incorporating Inorganic Fullerene-like nanoparticles. *Surface & Coatings Technology,* Vol.202, pp. 2287-2291
23. Johnson, B.B.; Santare, M.H., Novotny, J.E., Advani, S.G. (2009). Wear Behavior of Carbon Nanotube/High Density Polyethylene Composites. *Mechanics of Materials,* Vol.41, pp. 1108-1115
24. Karger-Kocsis, J.; Zhang, Z. (2005)Structure-property Relationships in Nanoparticle/semi- crystalline Thermoplastic Composites. In Balta Calleja, J.F., Michler, G., editors. In *Mechanical Properties of Polymers Based on Nanostructure and Morphology*. CRC Press, pp. 547-596, New York, USA
25. Kurokawa, M.; Uchiyama, Y., Nagai, S. (2000). Tribological properties and gear performance of polyoxymethylene composites. *Journal of Tribology, ASME,* Vol.122, pp. 809-814
26. Lee, J.Y.; Lim, D.P., Lim, D.S. (2007). Tribological Behaviour of PTFE Nanocomposite Films Reinforced with Carbon Nanoparticles. *Composites: Part B,* Vol.38, pp. 810-816
27. Li, F.; Hu, K., Li, J., Zhao, B. (2001). The Friction and Wear of Nanometer ZnO Filled Polytetrafluoroethylene. *Wear*, Vol.249, pp.877-882
28. Li, B.; Wood, W., Baker, L., Sui, G., Leer, C., Zhong, W.H. (2010). Effectual Dispersion of Carbon Nanofibers in Polyetherimide Composites and their Mechanical and Tribological Properties. *Polymer Engineering and Science,* Vol.50, pp. 1914-1922
29. Malucelli, G.; Palmero, P., Ronchetti, S., Delmastro, A., Montanaro,

M. (2010). Effect of Various Alumina Nanofillers on the Thermal and Mechanical Behavior of Low- density polyethylene-Al2O3 Composites. *Polymer International,* Vol.59, pp. 1084-1089

30. McCook, N.L.; Hamilton, M.A., Burris, D.L., Sawyer, W.G. (2007). Tribological Results of PEEK Nanocomposites in Dry Sliding Against 440C in Various Gas Environments. *Wear,* Vol.262, pp. 1511-1515
31. Misra, R.; Fu, X.B., Morgan, S.E. (2006). Surface Energetics, Dispersion, and Nanotribomechanical Behavior of POSS/PP Hybrid Nanocomposites. *Journal of Polymer Science: Part B: Polymer Physics,* Vol.45, pp. 2441-2455, DOI 10.1002/polb
32. Peng, Q.Y.; Cong, P.H., Liu, T.X., Huang, S., Li, T.S. (2009). The Preparation of PVDF/clay Nanocomposites and the Investigation of their Tribological Properties. *Wear,* Vol.266, pp. 713-720
33. Ravi Kumar, B.N.; Suresha, B., Venkataramareddy M. (2009). Effect of Particulate Fillers on Mechanical and Abrasive Wear Behaviour of Polyamide 66/polypropylene nanocomposites. *Materials and Design,* Vol.30, pp. 3852-3858
34. Sawyer, W.G.; Freudenberg, K., Bhimaraj, P., Schadler, L. (2003). A Study on the Friction and Wear Behavior of PTFE Filled with Alumina Nanoparticles. *Wear,* Vol.254, pp. 573-580
35. Shi, Y.; Feng, X., Wang, H., Lu, X., Shen, J. (2007). Tribological and Mechanical Properties of Carbon-Nanofiber-Filled Polytetrafluoroethylene Composites. *Journal of Applied Polymer Science,* Vol.104, pp. 2430-2437, DOI 10.1002/app.23951
36. Sirong, Y.; Zhongzhen, Y., Yiu-Wing M. (2007). Effects of SEBS-g-MA on Tribological Behavior of Nylon 66/organoclay Nanocomposites. *Tribology International,* Vol.40, pp. 855-862
37. Sun, L.H.; Yang, Z.G., Li, X.H. (2008). Mechanical and Tribological Properties of Polyoxymethylene Modified with Nanoparticles and Solid Lubricants. *Polymer Engineering and Science,* Vol.48, pp. 1824-1832
38. Sun, L.H.; Yang, Z.G., Li, X.H. (2008). Study on the Friction and Wear Behavior of POM/Al2O3 Nanocomposites. *Wear,* Vol.264, pp. 693-700
39. Vail, J.R.; Burris, D.L., Sawyer, W.G. (2009). Multifunctionality of Single-walled Carbon Nanotube-polytetrafluoroethylene nanocomposites. *Wear,* Vol.267, pp. 619-624
40. *Wang, Y.; Hu, X.G., Tian, M., Stengler, R. (2007). Study on Mechanical and Tribological Property of Nanometer ZrO2-filled Polyoxymethylene Composites.* Polymer-Plastics Technology and Engineering, *Vol.46, pp. 469-473*
41. Wang, Y.; Hu, K.H., Xu, Y.F., Hu, X.G. (2008). Structural, Thermal, and Tribological Properties of Intercalated Polyoxymethylene/molybdenum Disulfide Nanocomposites. *Journal of*

Applied Polymer Science, Vol.110, pp. 91-96

42. Wang, Y.; Lim, S., Luo, J.L., Zu, Z.H. (2006). Tribological and Corrosion Behaviour of Al2O3/polymer Nanocomposite Coatings. *Wear,* Vol.260, pp. 976-983
43. Werner, P.; Altstadt, V., Jaskula, R., Jacobs, O., Sandler, J.K.W., Shaffer, M.S.P., Windle, A.H. (2004). Tribological Behaviour of Carbon-nanofibre-reinforced Poly(ether ether ketone). *Wear,* Vol.257, pp. 1006-1014
44. Xian, G.; Zhang, Z., Friedrich, K. (2006). Tribological Properties of Micro- and Nanoparticles-Filled Poly(etherimide) Composites. *Journal of Applied Polymer Science,* Vol.101, pp. 1678-1686
45. Xiong, L.; Xiong, D., Yang, Y., Jin, J. (2011). Friction, Wear, and Tensile properties of vacuum hot pressing crosslinked UHMWPE/nano-HAP composites. *Journal of Biomedical Materials Research B: Applied Biomaterials,* Vol.98B, issue 1, pp. 127-138
46. Zhang, G. (2010). Structure-Tribological Property Relationship of Nanoparticles and Short Carbon Fibers Reinforced PEEK Hybrid Composites. *Journal of Polymer Science: Part B: Polymer Physics,* Vol.48, pp. 801-811
47. Zhang, G.; Chang, L., Schlarb, A.K. (2009). The roles of Nano-SiO2 particles on the tribological behaviour of short carbon fiber reinforced PEEK. *Composite Science and Technology,* Vol.69, pp. 1029-1035
48. Zhang, G.; Schlarb, A.K., Tria, S., Elkedim, O. (2008). Tensile and Tribological Behaviors of PEEK/nano-SiO2 Composites Compounded Using a Ball Milling Technique. *Composite Science and Technology,* Vol.68, pp. 3073-3080
49. Zhang, H.; Zhang, Z., Guo, F., Wang, K., Jiang, W. (2009). Enhanced Wear Properties of Hybrid PTFE/cotton Fabric Composites Filled with Functionalized Multi-walled Carbon Nanotubes. *Materials Chemistry and Physics,* Vol.116, pp. 183-190
50. Zhong, Y.J.; Xie, G.Y., Sui, G.X., Yang, R. (2011). Poly(ether ether ketone) Composites Reinforced by Short Carbon Fibers and Zirconium Dioxide Nanoparticles: Mechanical Properties and Sliding Wear Behavior with Water Lubrication. *Journal of Applied Polymer Science,* Vol.119, pp. 1711-1720
51. Zhou, Q.; Wang, K., Loo, L.S. (2009). Abrasion Studies of Nylon 6/Montmorillonite Nanocomposites Using Scanning Electron Microscopy, Fourier Transform Infrared Spectroscopy, and X-ray Photoelectron Spectroscopy. *Journal of Applied Polymer Science,* Vol.113, pp. 3286-3293

Chapter 12

Rubber Abrasion Resistance

Wanvimon Arayapranee

Rangsit University, Thailand

INTRODUCTION

Abrasion resistance is the ability of a material to resist mechanical action such as rubbing, scraping, or erosion that tends progressively to remove material from its surface. When a product has abrasion resistance, it will resist erosion caused by scraping, rubbing, and other types of mechanical wear. This allows the material to retain its integrity and hold its form. This can be important when the form of a material is critical to its function, as seen when moving parts are carefully machined for maximum efficiency. Abrasion resistant materials can be used for both moving and fixed parts in settings where wearing could become an issue.

The substances usually called "rubber" immediately brings to mind materials that are highly flexible and will snap back to their original shape after being stretched. In fact, there are three structural requirements for a given substance to be a rubber (i) rubber is made up of a polymer chain, liner or branched; (ii) the chain is flexible; and (iii) the chain is longer than a certain threshold length. Because the rubber is compliant and tough, it can easily absorb and survive a single strike of large deformation. However when used in contact with moving parts, a process of micro-tearing can occur on the rubber surface around the sharp asperities, gradually removing the material and finally terminating the functional life of the rubber. In many applications, abrasive wear is the major failure mode of rubbers. In normal materials, a rough surface is made smooth by repeated friction or abrasion with harder materials. Rubber hardly ever slides on other rubber like materials but on tracks grossly dissimilar from it in surface texture, chemical constitution and elastic behavior, however, when the smooth surface is abraded, periodic parallel ridged patterns, looking like a wind-wrought pattern on sand, are formed on the rubber surface. These typical patterns are held through all processes of rubber abrasion, on the surface of tires, conveyor belts, printing rolls and shoes for example, which are thus regarded as the essential basis of rubber abrasion.

Abrasion process involves removal of small particles (1-5 □m) leaving behind pits in the surface and then followed by removal of large particles (> 5 □m) (El-Tayeb & Nasir, 2007). Detachment of small particles plays an important role in initiating the abrasion (Muhr & Roberts, 1992) and this is related to either a structural unit or localized stresses in the rubber. Since abrasion is clearly a manifestation of mechanical failure, Shallamach (Schallamach, 1557/58, 1968) used tearing energy to describe the rubber wear mechanisms, and Ratner et al. (Ratner et al., 1967) has established an equation in which the wear loss is related to macroscopic mechanical properties such as tensile strength, elongation at break, hardness etc. Thomas (Thomas, 1958, 1974) proposed the problem of abrasion is presented using fracture mechanics which treats fatigue and tensile failure as crack growth processes from small flaws. Crack growth can be influenced by the presence of oxygen or ozone. The nature of the vulcanizing system affects strength: crosslinks probably rupture and reform under stress. Other suggested factors responsible for particle detachment are internal subsurface failure due to flaws in the rubber such as dirt or voids (Gent, 1989) or interfacial adhesion at high speed rolling (Roberts, 1988). According to Pandey et al. (Pandey et al., 2003), there is no distinction between wear and abrasion, although other researcher (Schallamach, 1957/58) defined abrasion as that produced by laboratory machine on a rubber piece and wear as something that happens to tires or other rubber products. Thus, for rubber, "abrasion" covers all mechanisms, whereas the word "abrasion" for other materials refers in particular to scoring by hard, sharp particles. In the absence of transient effects such as clogging of the abrasive or evolution of an abrasion pattern it is found that the quantity of rubber abraded is proportional to the distance of sliding between rubber and counterface. However, wear of tires and abrasion on certain laboratory abrasion machines (e.g. the Akron abrader) brings into play gross properties of tire or test piece which affect the rate of wear just as much as does the abrasion resistance of the compound. However, Muhr and Roberts (Muhr & Roberts, 1992) proposed that abrasion reserves for processes where the amount of sliding is controlled, and that wear applies to the many practical situations for which the amount of sliding is as significant a variable as the attrition per unit sliding distance.

Schallamach (Schallamach, 1957/58, 1968) defines abrasion of rubber as a purely mechanical failure produced frictionally by the asperities of the track and this process creates periodic structures often called "abrasion pattern", a series of parallel ridges perpendicular to the sliding direction created on the surface of rubber during abrasive wear. He proposed the mechanism of rubber abrasion from a fracture mechanics point of view, relating the rate of wear to the crack growth resistance of the rubber. Although the concept of crack growth plays a very important role in abrasion, particularly in the growth of a single ridge, when we consider that the essential subject of rubber abrasion is in the formation of the periodic surface pattern consisting of very many cracks, moreover abrasive wear is a consequence of friction, in other words, it is impossible for any abrasive transfer of material to occur without friction phenomena. As is well known, an abrasion pattern is formed at the initial stage of abrasion and grows in ridge spacing and ridge height, whose geometric feature remains constant in appearance once it has grown up to the critical size. The abrasion pattern moves very slowly along in the sliding direction in a manner that the crack at the root of the pattern wedge is deepened somewhat and the protruding flap is tom off.

When rubber is slided over another abrasive surface, contact of abrasive grits (asperities) occurs. Upon application of a normal load, the extremely low tensile modulus of rubber ensures extensive deformation to establish a conformal contact with the counter part. This causes the real area of contact to become comparable to the apparent area of contact. Owing to the curved and entangled structure of chains of molecules in rubber, they can undergo considerable lateral deformation without fracture by stretching and twisting of chains. Gent (Gent, 1989) proposed a hypothetical mechanism for creating subsurface cracks during frictional sliding as part of the process of abrasive wear of rubber. It consists of the unbounded elastic expansion of microscopic precursor voids until they burst open as cracks, under the action of internal pressure or of a triaxial tension in the surrounding rubber. This conjecture accounts for enhanced resistance to abrasion for compounds reinforced with carbon black, in terms of increased stiffness without much loss of extensibility, and for the lack of correlation of abrasion resistance with other measures of strength. It should be noted that it is specific to soft, extensible materials, and thus it also accounts for marked differences in the nature of the wear process in rubbery materials compared to plastics and metals. Only rubbery materials appear to abrade away by a linking up of microcracks at right angles to the sliding direction to produce characteristic wear ridges known as the Schallamach abrasion pattern. Three mechanisms of generating a sufficiently large inflation pressure or triaxial tension are discussed. The most probable one seems to be thermal decomposition of rubber, generating volatile decomposition products a microscale blowout process. This would be aggravated by a simultaneous softening of the rubber on heating. Although strictly conjectural, it would be helpful to know whether the character of wear changes when abrasion is carried out under a large superimposed hydrostatic pressure. Sliding contact generates a shear stress along the surface of rubber and extensive shear stress along the stress axis is found to occur owing to the nature of elastomer (Chandrasekaran & Batchelor, 1997). When the shear stress exceeds the cohesive strength of the chain, fracture occurs by propagation of crack along the root of the contact area. The sudden release of deformation energy in the form of fracture in the contact area surface results in recovery of rubber to original coiled and entangled state. The visco-elastic nature of rubber limits the rate of recovery which results in shear wave propagation along the surface of rubber during sliding. Fukahori and Yamazaki (Fukahori & Yamazaki, 1994) proposed a new concept to understand the mechanism of formation of the periodic patterns characteristic in rubber abrasion. They showed that the driving force to generate the periodic surface patterns, and thus rubber abrasion consist of two kinds of periodic motion, stick-slip oscillation and the microvibration generated during frictional sliding of rubber. The stick-slip oscillation is the driving force to propagate cracks, then abrasion patterns and the microvibration with the natural frequency of the rubber induced in the slip phase of the stick-slip oscillation is another driving force for the initiation of the cracks. Although initial cracks originate in the slip region of the rubber surface, the propagation of the cracks is strongly excited in the stick region. Accordingly, the initial size of the abrasion pattern, pattern spacing, equals the distance determined by the natural period of the rubber and the sliding velocity while the constant pattern spacing after the critical number of frictional slidings agrees with the distance given by the period of the stick-slip oscillation and the sliding velocity. Consequently, during rubber abrasion, two driving forces produce bimodal size distribution of abraded particles, small particles of the order of ten micrometres by microvibrations and large ones of the order of a few hundred micrometres by the stick-slip motions. Sliding of rubber with high frictional forces does not necessarily entail abrasion (as it does for metal-metal contacts). Rather, abrasion of rubber results from mechanical failure due to excessively high local frictional stresses which are most likely to occur on rough tracks.

In spite of its practical importance, abrasion is perhaps the least understood phenomena amongst the various types of failures of rubber, as it is difficult to predict the abrasion behaviour from other rubber properties (Pandey et al., 2003). It is influenced by the hysteresis properties of the vulcanizates, magnitude of the frictional force and the resistance of rubber to rupture. The abrasion is a combination of mechanical, mechanochemical and thermochemical processes. The formation of an abrasion pattern depends on several factors such as crack growth process (Uchiyama and Ishino, 1992), mechanical properties of the rubber, and on the chemical, ageing and thermal conductivity properties of the composite. It depends on the modulus of elastomer, the abrasion pattern may be characterized by either ridges at the right angles to the rubbing direction if the modulus of elastomer is sufficiently low or score lines parallel to the rubbing direction if the modulus of elastomer is sufficiently high (Evstratov et al., 1967). There are a number of ways to make a material resistant to abrasion. One option is to utilize a special coating which creates a hardened layer over the material and resists friction. Some materials are also naturally extremely hard, and are ideal for settings in which abrasion resistance will be desirable. Other materials can be specifically formulated for increased hardness, as seen in plastics facilities which manufacture abrasion resistant plastics with the use of chemicals which harden and strengthen the plastic. The properties of a vulcanized rubber can be significantly influenced by details of the compounding. Practical materials will have, in addition to the base polymer, fillers, anti- degradants, crosslinking agents, accelerators etc. All of these can have an influence on the physical and chemical stability of the finished material.

RUBBER USE

Compounding

The rubber industry began when Charles Goodyear developed the first useful rubber compound: natural rubber plus sulfur. The concept of mixing materials into rubber to improve performance is still primary importance today. Without compounding, few rubbers would be of any commercial value. Any given rubber application will have a long list of necessary criteria in addition to cost, encompassing appearance, processing, mechanical, electrical, chemical, and thermal properties. Developing such compounds requires a broad knowledge of material science and chemistry combined with experience. The use of designed experiments can greatly facilitate selecting the optimum compound formulation.

The major components in a compound are rubber, vulcanizing agents, fillers, plasticizers, and antidegradants.

Rubber

Rubber is polymeric material endowed with the properties of flexibility and extensibility: with the application of force, the molecules straighten out in the direction in which they are being pulled; on release from being extended, they spontaneously recover their normal, random arrangements. Rubbers include natural rubber, a naturally occurring substance obtained from the exudations of certain tropical plants, and synthetic rubbers, artificially derived from petrochemical products. Among the most important synthetics are styrene-butadiene, polybutadiene, and polyisoprene (commonly classified as the "general purpose"), as well as ethylene-propylene rubber (often referred to as "specialty rubber"). The prices of these synthetics have been historically in the range of natural rubber prices and their markets have, although to varying degrees, overlapped those of natural rubber. Synthetic rubbers are materials with distinctive chemical structures, whereas the emphasis with natural rubber lies on different types and grades within one single broad category.

Natural rubber

Natural rubber (polymer designation cis-1-4 polyisoprene, empirical formula (C5H8)n obtained from the sap (latex) of several rubber-yielding plants (e.g. Hevea Brasiliensis and Parthenia argentatum) by coagulation with chemicals, drying, electrical coagulation, and other processes is the prototype of all elastomers. Latex extracted in the form of latex from the bark of the Hevea tree is a polydispersed colloidal system of rubber particles in an aqueous phase. With Hevea latex, the dry rubber content varies between approximately 28- 40%, although it may rise to 45-50% after a long period of non-tapping (notice that, for statistical purposes, figures for natural rubber may include the dry rubber content of latex). The rubber produced from latex contains, besides the hydrocarbon, relatively small quantities of protein, carbohydrates, resin-like substances, mineral salts, and fatty acids.

The natural rubber polymer is nearly 100% cis-1,4 polyisoprene with Mw ranging from 1 to

2.5×10^6. Due to its high structural regularity, natural rubber tends to crystallize spontaneously at low temperatures or when it is stretched. Low temperature crystallization causes stiffening, but is easily reversed by warming. The strain-induced crystallization gives natural rubber products outstanding resilience, flexibility, and tear and tensile strength, as well as low heat build-up and abrasion. However, a drawback is natural rubber moderate environmental resistance to factors such as oxidation and ozone; so too for its scarce resistance to chemicals, including gasoline, kerosene, hydraulic fluids, degreasers, synthetic lubricants, and solvents. In addition, latex contains proteins that can cause severe allergic response in a small percentage of the population and among medical professionals following extensive exposure. The largest use of natural rubber is in the tires. Over 70 percent of its manufacture consumption is in this area. The next largest use is as latex in dipped goods, adhesives, rubber thread, and foam. These uses account for approximately another 10 percent. The of applications remainder is used in a variety such as convey or belts, hoses, gaskets, footwear, and antivibration devices such as engine mounts.

Polyisoprene

Synthetic polyisoprene is designed to be similar to natural rubber in structure and properties. Although it still demonstrates lower green strength, slower cure rates, lower hot tear, and lower aged properties than its natural counterpart, synthetic polyisoprene exceeds the natural types in consistency of product, cure rate, processing, and purity. In addition, it is superior in mixing, extrusion, molding, and calendering processes. The lithium based polymers were found to produce up to 94 percent cis, which still was not high enough to provide the properties of natural rubber. Polymers made with the coordination catalysts have cis contents of up to 98 percent, providing products that can more closely serve as replacements for natural rubber than the lithium-based polymers. In comparison with natural rubber, they offer the advantage of a more highly pure rubber (no non-rubber material) and excellent uniformity. A high trans-1,4 structure was produced by Polysar, and is now being produced by Kuraray. A Li-IR with increased 3,4 structure can be prepared by adding polar modifiers to the alkyl lithium catalyst system. However, since the higher cis- 1,4 configuration most closely mirrors the properties of natural rubber and is the most important commercially. Currently synthetic polyisoprene is being used in a wide variety of industries in applications requiring low water swell, high gum tensile strength, good resilience, high hot tensile, and good tack. Gum compounds based on synthetic polyisoprene are being used in rubber bands, cut thread, baby bottle nipples, and extruded hose. Carbon black loaded compounds find use in tires, motor mounts, pipe gaskets, shock absorber bushings and many other molded and mechanical goods. Mineral filled systems find applications in footwear, sponge, and sporting goods. In addition, recent concerns about allergic reactions to proteins present in natural rubber have prompted increased usage of the more pure synthetic polyisoprene in some applications.

Styrene-butadiene rubber

The largest-volume synthetic rubber consumed is styrene-butadiene rubber (SBR) produced by both emulsion (E-SBR) and solution (S-SBR). In 2003, SBR solid rubber accounted for 4% percent of all synthetic rubber. If SBR latex and carboxylated SBR latex are included, its share increases to 55 percent. The major application of solid SBR is in the automotive and tire industry, accounting for approximately 70 percent of the use. Therefore, SBR has been tightly tied to the tire business.

Most of the E-SBR contains about 24% by weight of styrene and it is a random copolymer with butadiene. Some specific grades contain as much as 40-46% styrene, and are much stiffer. The polymerization is by free radical initiator and there is a finite probability of chain-transfer reaction, which generates long branches.

Emulsion polymerization is carried out either hot, at about 25-50□C, or cold, at about 5-25□C, depending upon the initiating system used. SBR made in emulsion usually contains about 24% styrene randomly dispersed with butadiene in the polymer chains. At high temperature polymerization, many long-braches and gels were formed. The rubber was stiff and difficult to mill, mix, or calender than natural rubber, deficient in building tack, and having relatively poor inherent physical properties. Processability and physical properties were found to be greatly improved by the addition of process oil and reinforcing pigments. Polymerization at a lower temperature became possible, giving less branches and gels. "Cold" SBR generally has a higher average molecular weight and narrower molecular weight distribution. It thereby offers better abrasion and wear resistance plus greater tensile and modulus than "hot" SBR. Since higher molecular weight can make cold SBR more difficult to process, it is commonly offered in oil-extended form. S-SBR comes in two distinctly different subgroups, one made by an anionic initiator and the other by free radical initiators. SBR made in solution contains about the same amount of styrene, but both random and block copolymers can be made. Solution SBR can be tailored in polymer structure and properties to a much greater degree than their emulsion counterparts. The random copolymers offer narrower molecular weight distribution, low chain branching, and lighter color than emulsion SBR. They are comparable in tensile, modulus, and elongation, but offer lower heat buildup, better flex, and higher resilience. Certain grades of solution SBR even address the polymer's characteristic lack of building tack, although it is still inferior to that of natural rubber. The processing of SBR compounds in general is similar to that of natural rubber in the procedures and additives used. SBR is typically compounded with better abrasion, crack initiation, and heat resistance than natural rubber. SBR extrusions are smoother and maintain their shape better than those of natural rubber. SBR was originally developed as a general purpose elastomer and it still retains this distinction. It is the largest volume and most widely used elastomer worldwide. Its single largest application is in passenger car tires particularly in tread compounds for superior traction and tread wear. Substantial quantities are also used in footwear, foamed products, wire and cable jacketing, belting, hoses, and mechanical goods.

Polybutadiene rubber

Polybutadiene rubber (BR) was originally made by emulsion polymerization, generally with poor results. It was difficult to process and did not extrude well. This rubber became commercially successful only after it was made by solution polymerization using stereospecific Ziegler-Natta catalysts.

This provided a polymer with greater than 90% cis- 1,4-polybutadiene configuration. This structure hardens at much lower temperatures (with Tg of -100□C) than natural rubber and most other commercial rubbers. This gives better low temperature flexibility and higher resilience at ambient temperatures than most rubbers. Greater resilience means less heat buildup under continuous dynamic deformation as well. This high-cis BR was also found to possess superior abrasion resistance and a great tolerance for high levels of extender oil and carbon black. High-cis BR was originally blended with natural rubber simply to improve the latter's processing properties, but it was found that the BR conferred many of its desirable properties to the blend. The same was found to be true in blends with SBR.

The 1,3-butadiene monomer can polymerize in three isomeric forms: by cis 1,4 addition, trans 1,4 addition, and 1,2 addition leaving a pendant vinyl group. By selection of catalyst and control of processing conditions, polybutadiene is now sold with various distributions of each isomer within the polymer chain, and with varying levels of chain linearity, molecular weight and molecular weight distribution. Each combination of chemical properties is designed to enhance one or more of BR's primary attributes.

The major use of polybutadiene (cis-1,4-BR) having a very low glass transition temperature in the region -75□C to -100□C is in tires, with over 70 percent of the polymer produced used by the tire industry, primarily in blends with SBR or natural rubber to improve hysteresis (resistance to heat buildup), abrasion resistance, and cut growth resistance of tire treads. The type of BR used depends on which properties are most important to the particular compound. High-cis and medium-cis BR have excellent abrasion resistance, low rolling resistance, but poor wet traction. High-vinyl BR offer good wet traction and low rolling resistance, but poor abrasion resistance. Medium-vinyl BRs balance reasonable wet traction with good abrasion resistance and low rolling resistance. Polybutadiene is also used for improved durability and abrasion and flex crack resistance in tire chaffer, sidewalls and carcasses, as well as in rubber blends for belting. High- and medium-cis BRs are also used in the manufacture of high impact polystyrene. Three to twelve percent BR is grafted onto the styrene chain as it polymerizes, conferring high impact strength to the resultant polymer.

Polybutadiene made by emulsion polymerization with a free radical initiator is used as the rubber component of an impact modifier in plastics, in particular high impact polystyrene (HIPS) and acrylonitrile-butadiene-styrene resin (ABS). In the HIPS application the rubber is dissolved in the styrene monomer, which is then polymerized via a free-radical mechanism. A complex series of phase changes occurs, resulting in small rubber particles containing even smaller polystyrene particles being incorporated into a polystyrene matrix. The rubber is added to increase impact strength. Because of the unique morphology that is formed, low levels of rubber (typically around 7%) provide rubbery particles having a volume fraction of 30-40 percent. This morphology leads to high impact at very low rubber levels, providing good stiffness and hardness.

There is also a fairly large market for high cis BR in solid core golf balls. In this application, the polymer is compounded with zinc acrylate and the mixture is cured with peroxide. This produces an ionically cross-linked compound that has outstanding resilience. The covers are also ionomers with superior cut resistance. In the last few years the golf ball market has been shifting away from the traditional wound ball to these new solid core balls that use polybutadiene.

Nitrile rubber

Acrylonitrile butadiene rubber (NBR) is made as an emulsion with a free radical initiator. Polymers are made with an acrylonitrile (AN) content of, for example, 28, 33 or 40 weight percent, depending upon the required oil resistance. It also has good elongation properties as well as adequate resilience, tensile and compression set. The major applications for this material are in areas requiring oil and solvent resistance. As the acrylonitrile content increases in the polymer chain, the properties change predictably. The glass transition temperature increases approximately 1.5□C for each percent increase in acrylonitrile. Properties such as hysteresis loss, resilience, and low-temperature flexibility will correspondingly change. The oil resistance increases with increased acrylonitrile content, as does the compatibility with polar plastics such as PVC. The major market for nitrile rubber is in the automotive area because of its solvent and oil resistance. Major end uses are for hoses, fuel lines, O-rings, gaskets, and seals. In blends with polyvinyl chloride (PVC) and acrylonitrile butadiene styrene (ABS), nitrile rubber acts as an impact modifier. Some nitrile rubber is sold in latex form for the production of grease-resistant tapes, gasketing material, and abrasive papers. Latex also is used to produce solvent resistant gloves.

Hydrogenated NBR (HNBR) is produced by first making an emulsion-polymerized NBR using standard technique. Almost all the butadiene units become saturated to produce an ethylene-butadiene-acrylonitrile terpolymer. These "post-polymerization" reactions are very expensive so HNBR usually command a premium price. HNBR is usually cured with peroxides, similar to ethylene-propylene rubber, because it has no unsaturation for a conventional sulfur cure system. HNBR has many uses in the oil-field, including down hole packers and blow-out preventers, because of its outstanding oil resistance and thermal stability. For the same reasons, it has also found uses in various automotive seals, O-rings, timing belts, and gaskets. Resistance to gasoline and aging make HNBR ideal for fuel-line hose, fuel-pump and fuel injection components, diaphragms, as well as emission-control systems. HNBR is the best selection to achieve to the highest abrasion and heat resistance. Service temperature of this rubber is up to 160□C and it's used in temperatures up to 200□C for short times and its abrasion resistance and wet traction is very good. But price of HNBR is high and it isn't an economy rubber for general applications.

Ethylene-propylene rubber

Ethylene-propylene rubber continues to be one of the most widely used and fastest growing synthetic rubber having both specialty and general purpose applications. Polymerization and catalyst technologies in use today provide the ability to design polymers to meet specific and demanding application and processing needs. Versatility in polymer design and performance has resulted in broad usage in automotive weather-stripping and seals, glass run channel, radiator, garden and appliance hose, tubing, belts, electrical insulation, roofing membrane, rubber mechanical goods, plastic impact modification, thermoplastic vulcanizates and motor oil additive applications. Ethylene-propylene rubber are valuable for their excellent resistance to heat, oxidation, ozone and weather aging due to their stable, saturated polymer backbone structure. Properly pigmented black and non-black compounds are color stable. As non-polar rubber, they have good electrical resistivity, as well as resistance to polar solvents, such as water, acids, alkalies, phosphate esters and many ketones and alcohols. Amorphous or low crystalline grades have excellent low temperature flexibility with glass transition points of about minus 60°C. Ethylene-propylene rubber uses the same chemical building blocks or monomers as polyethylene (PE) and polypropylene (PP) thermoplastic polymers. These ethylene (C2) and propylene (C3) monomers are combined in a random manner to produce rubbery and stable polymers. There are two general types of polymers based on ethylene and propylene: ethylene- propylene rubber (EPM) and ethylene-propylene terpolymer (EPDM). EPM accounts for approximately 20 percent of the polyolefin rubber produced. Comprising a totally saturated polymer, these materials require free-radical sources to cross-link. EPDM was developed to overcome this cure limitation. For EPDM a small amount (less than 15%) of a nonconjugated diene is terpolymerized into the polymer. One of the olefinic groups is incorporated into the chain, leaving its other unsaturated site free for vulcanization or polymer modification chemistry. This ensures that the polymer backbone remains saturated, with corresponding stability, while still providing the reactive side group necessary for conventional cure systems. The nonconjugated dienes used commercially are ethylidene norbornene, 1,4 hexadiene, and dicyclopentadiene. Each diene incorporates with a different tendency for introducing long chain branching (LCB) or polymer side chains that influence processing and rates of vulcanization by sulfur or peroxide cures.

Vulcanizing agents

Vulcanization is a chemical process for converting rubber or related polymers into more durable materials via the addition of sulfur or other equivalent vulcanizing agent. The function of vulcanizing agent is to modify the polymer by forming crosslinks (bridges) between individual polymer chains; the most common ones are the sulfur type for unsaturated rubber and peroxides for saturated polymers. Uncured natural rubber is sticky, deforms easily when warm, and is brittle when cold. In this state, it is a poor material when a high level of elasticity is required. Vulcanized material is less sticky and has superior mechanical properties. A vast array of products is made with vulcanized rubber including tires, shoe soles, hoses, and hockey pucks. The main polymers subjected to vulcanization are polyisoprene (natural rubber) and styrene-butadiene rubber, which are used for most passenger tires. Chemicals called accelerators may be added to control the cure rate in the sulfur system; these materials generally are complex organic chemicals containing sulfur and nitrogen atoms. Stearic acid and zinc oxide usually are added to activate these accelerators. Metal oxides are used to cure halogenated polymers such as polychloroprene or chlorosulfonated polyethylene.

Fillers

Natural and synthetic rubbers, also called elastomers are rarely applied in their pure form. They are too weak to fulfill practical requirements because of lack of hardness, strength properties and wear resistance. A rubber compound contains, on average, less than 5 kg of chemical additives per 100 kg of rubber, while filler loading is typically 10-15 times higher. Of the ingredients used to modify the properties of rubber products, the filler often plays a significant role. Most of the rubber fillers used today offer some functional benefit that contributes to the processability or utility of the rubber product. Styrene butadiene rubber, for example, has virtually no commercial use as an unfilled compound. Fillers are used in order to improve the properties of rubber compounds. The characteristics which determine the properties a filler will impart to a rubber compound are particle size, particle surface area, particle surface activity and particle shape. Surface activity relates to the compatibility of the filler with a specific rubber and the ability of the rubber to adhere to the filler. Rubber articles derive many of their mechanical properties from the admixture of these reinforcing (active) fillers at quantities of 30% up to as much as 300% relative to the rubber part. The introduction of carbon black as a reinforcing agent in 1904, lead to strongly increased tread wears resistance. Carbon black is in use as the most versatile reinforcing filler for rubber, complemented by silicas. In tire manufacturing silicas are more and more used nowadays, mainly to decrease the rolling resistance. The increased attitude of protecting the environment gives rise to a demand for tires combining a long service life with driving safety and low fuel consumption, achieved by this lower rolling resistance. However, the change from carbon black to silica is not at all obvious because of technical problems involved. In particular, the mixing of rubber with pure silicas is difficult, because of the polarity-difference between silica and rubber. Therefore, coupling agents are applied in order to bridge this polarity difference. Sometimes fillers are added to reduce cost, increase hardness, and color the compound. Generally they do not provide the dramatic improvement in properties seen with reinforcing agents, but they may have some reinforcing capability. Carbon black and silica are the most common reinforcing agents. These materials improve properties such as tensile strength and tear strength; also, they increase hardness, stiffness, and density and reduce cost. Almost all rubbers require reinforcement to obtain acceptable use properties. The size of the particles, how they may be interconnected (structure), and the chemical activity of the surface are all critical properties for reinforcing agents. In tire applications, new polymers are currently being developed which contain functional groups that directly interact with carbon black and silica, improving many properties. Typical fillers are clays, calcium carbonate, and titanium dioxide.

Plasticizers

These materials are added to reduce the hardness of the compound and can reduce the viscosity of the uncured compound to facilitate processes such as mixing and extruding. The most common materials are petroleum-based oils, esters, and fatty acids. Critical properties of these materials are their compatibility with the rubber and their viscosity. Failure to obtain sufficient compatibility will cause the plasticizer to diffuse out of the compound. The oils are classified as aromatic, naphthenic, or paraffinic according to their components. Aromatic oils will be more compatible with styrene-butadiene rubber than paraffinic oils, whereas the inverse will be true for butyl rubber. The aromatic oils are dark colored and thus cannot be used where color is critical, as in the white sidewall of a tire. The naphthenic and paraffinic oils can be colorless and are referred to as nonstaining.

Antidegradents

An antidegradant, this group of chemicals is an ingredient in rubber compounds to deter the aging of rubber products. The most important are the antioxidants, which trap free radicals and prevent chain scission and crosslinking. Antiozonants are added to prevent ozone attack on the rubber, which can lead to the formation and growth of cracks. Antiozonants function by diffusion of the material to the surface of the rubber, thereby providing a protective film. Certain antioxidants have this characteristic, and waxes also are used for this purpose.

Processing

A wide range of processes are used to convert a bale of rubber into a rubber product such as a tire. The first process generally will be compounding. Typical compounding ingredients were discussed previously. In many compounds more than one rubber may be needed to obtain the performance required. Uncured rubber can be considered as a very high viscosity liquid; it really is a viscoelastic material possessing both liquid and elastic properties. Mixing materials into rubber requires high shear, and the simplest method is a double roll mill in which the rubber is shear-mixed along with the other compounding ingredients in the bite of the mill. Large scale mixing is most commonly done with a high-shear internal mixer called a Banbury. This mixing is a batch process, although continuous internal mixers also are used. The compounded rubber stock will be further processed for use. The process could be injection or transfer molding into a hot mold where it is cured. Tire curing bladders are made in this fashion. Extrusion of the rubber stock is used to make hose or tire treads and sidewalls. Another common process is calendaring, in which a fabric is passed through rolls where rubber is squeezed into the fabric to make fabric-reinforced rubber sheets for roofing membranes or body plies for tires. The actual construction of the final product can be quite complex. For example, a tire contains many different rubber components some of which are cord or fabric reinforced. All of the components must be assembled with high precision so that the final cured product can operate smoothly at high speeds and last over 50,000 miles.

ABRASION TEST

An abrasion test is a test used to measure the resistance of a material to wear stemming from sliding contact such as rubbing, grinding, or scraping against another material. Abrasion may be measured in a variety of ways, depending on the resistance test used and the information that is desired from the test (Dick, 2001). For example, where the amount of material lost is a concern regardless of whether the material fails, abrasion may be measured in terms of the percentage of material lost, either by mass or by volume, between the start and end of the test. Another measure sometimes used is the number of abrasion cycles a material withstands before failure.

This would be more appropriate if information on how long the material or product will survive before outright failure is of primary interest. Abrasion tests try to accelerate the process by applying more cutting-like conditions; however, this approach may not simulate actual wear. It is also important to try to match the severity of the abrasion test to the severity of the product wear conditions. For example, the severity of test conditions imparted by most abraders is usually greater than what the highway pavement may impart to a tire tread compound during normal driving. Wear resistance is an important rubber compound property related to the useful product life for tires, belts, shoe soles, rubber rolls, and sandblasting hose, among other products. A wide variety of different abrasion testers have been developed over the years in an attempt to correlate to these product wear properties. Several factors are typically considered in developing or selecting an appropriate abrasion test for the application at hand. The shape of the contact area is taken into consideration, as is the composition of the two surfaces making contact with one another. Speed of sliding contact between the two surfaces, the force with which they act on one another, and the duration of contact between them may also be considered. In addition to the materials themselves, the environment in which they are making contact also plays a role in selecting an appropriate abrasion test.

The abrasion resistance is expressed as volume loss in cubic millimetres or abrasion resistance index in percent. For volume loss, a smaller number indicates better abrasion resistance, while for the abrasion resistance index; a smaller number denotes poorer abrasion resistance. Tested compounds are usually compared on a "volume loss" basis which is calculated from the weight loss and density of the compound. Abrasion test results are known to be variable; therefore, it is important to control and standardize the abradant used in the test. Also, it is a good idea to relate test results to a standard reference vulcanizates.

ASTM D394, the Dupont Abrasion Test Method, consists of a pair of rubber test pieces pressed against a disk of a specified abrasive paper which rotates whilst a pair of moulded test pieces is continuously pressed against it either with a constant force or with a force adjusted to give a constant torque on the arm holding the test pieces. Care should be taken with soft rubber compounds because "smearing" can occur, affecting test results.

ASTM D1630 describes the rotary-platform, double-head abrader is commonly referred to as the NBS Abrader used on rubber compounds for shoe soles and heels. The NBS abrader uses rotating drums with a specified abrasive paper around them onto which the test pieces are pressed by means of levers and weights, a specified standard reference compound to be used for the calculation of an abrasive index.

ASTM D2228 describes the Pico Abrader. This unique test works on the principle of abrading the rubber surface by rotating a rubber specimen against a pair of tungsten carbide knives. A special dusting power is fed to the test piece surface, which doubtless helps to avoid stickiness. This method specifies five standard rubbers and the result also expressed as an abrasion index. Force on the test piece and speed of rotation can be varied and, presumably, different abradant geometries could be used, although the distinctive feature of the Pico is the use of blunt metal knives in the presence of a powder.

ASTM D3389 refers to the Taber Abrader using a pair of abrasive wheels, a method not originally from the rubber industry. This very general method uses two abrasive wheels against the rubber test piece (disk) attached to a rotating platform. Although the degree of slip cannot be varied; however, the force on the test piece and the nature of the abradant are very readily varied and tests can be carried out in the presence of liquid or powder lubricants. When using the usual type of abrasive wheel, a refacing procedure is carried out before each material tested.

ISO 4649 refers to the DIN Abrader, based on the German Standard. The rubber test piece with a holder is traversed a rotating cylinder covered with a sheet of the abradant paper. By allowing the sample holder to move the test piece across the drum as it rotates, there is less chance of rubber buildup on the abradant paper. This method, used extensively in Europe, is very convenient and rapid and well suited to quality control the uniformity of a specific material. The achieved test results provide important parameters in respect to the wear of rubbers in practical use. The details of procedure and expression of results are something of a compromise, being a compilation of the German approach and the British approach. Two procedures are specified, using a rotating or non-rotating test piece respectively. In principle, the abrasion should be more uniform if the test piece is rotated during test. The standard abradant is specified in terms of weight loss of a standard rubber using a non- rotating test piece and has to be run in against a steel test piece before use. Results can either be expressed as a relative volume loss with the abradant normalized relative to a standard rubber or as an abrasion index relative to a standard rubber.

British Standard BS903: Part A9 still describes the Akron Abrader. The rubber test piece is a moulded wheel which is positioned against an abrasive cylinder under constant speeds and held against the abrasive wheel by a constant force. The Akron Abrader has the advantage of allowing variation in the degree of slip in the test by varying the angle of the test piece.

EFFECT OF COMPOUNDING INGREDIENTS ON ABRASION RESISTANCE

Rubber

In rubbery materials, when the smooth surface is abraded, periodic parallel ridged patterns are formed on the rubber surface. These typical patterns are held through all processes of rubber abrasion, on the surface of tires, conveyor belts, printing rolls and shoes for example, which are thus regarded as the essential basis of rubber abrasion. In the absence of any serious chemical decomposition the abrasion process initially results in the removal of small rubber particles just a few microns in size, leaving pits behind in the surface. With continued rubbing, larger pieces of rubber are removed. Although most weight loss is attributable to the larger pieces, it is thought that the detachment of the smaller particles initiates the abrasion process. The small particles have a characteristic size of 1-5 □m, but whether this relates to a structural unit in the rubber compound (Muhr & Roberts, 1992). Other suggestions are that mechanical rupture to produce the particles relates to flaws in the rubber, including dirt, or voids that cavitate leading to internal subsurface failure (Gent, 1989). A rolling experiment suggested that particle detachment might be linked to interfacial adhesion (Roberts, 1988). Schallamach (Schallamach, 1557/58, 1968) reported that rubber often develops a pattern of ridges perpendicular to the direction of abrasion. In the simplest case abrasion is produced by a line contact pulling a tongue of rubber from the ridge producing crack growth at the base of the tongue. Provided the surface configuration is in a steady state, the quantity of rubber abraded can be related quantitatively to the frictional force and the crack growth characteristic of the rubber. The abrasion of rubber results from mechanical failure due to excessively high local frictional stresses which are most likely to occur on rough tracks. Theories of abrasion thus require details of the local stresses, which together with the strength properties of the rubber may enable the rate of abrasion to be predicted. Gent and Pulford (Gent & Pulford, 1983) reported the reversal in the relative rates of wear of unfilled polybutadiene rubber comparing to those of unfilled natural rubber and styrene butadiene rubbers as frictional force increased. Fukahori and Yamazaki (Fukahori & Yamazaki, 1994) investigated the mechanism of the formation of the periodic ridges in rubber abrasion by designing the razor blade abrader. They reported that the driving force to generate the periodic surface patterns, and thus rubber abrasion consists of two kinds of periodic motions, stick-slip oscillation and the microvibration generated during frictional sliding of rubber. The stick-slip oscillation is the driving force to propagate cracks, then abrasion patterns and the microvibration with the natural frequency of the rubber induced in the slip phase of the stick-slip oscillation is another driving force for the initiation of the cracks. Although initial cracks originate in the slip region of the rubber surface, the propagation of the cracks is strongly excited in the stick region. Champ et al. (Champ et al., 1974) proposed the mechanism of rubber abrasion from a fracture mechanics point of view, relating the rate of wear to the crack growth resistance of the rubber. Although the concept of crack growth plays a very important role in abrasion, particularly in the growth of a single ridge, when consider that the essential subject of rubber abrasion. Liang et al. (Liang et al., 2010) investigated the blade abrasion of four different rubber materials, unfilled natural rubber, unfilled styrene butadiene rubber, unfilled polybutadiene rubber and carbon black filled styrene-butadiene rubber. Each is abraded until the steady state abrasion pattern is developed on the surface of moulded rubber wheels. The steady state conditions are measured using the weight loss per revolution of the wheel. The abraded surface is cut to examine the typical asperity profile. Each profile is modeled using finite element analysis to calculate the stored energy release rate for each combination of material and test condition. The stored energy release rate when combined with an independent measure of the rate of crack growth measured using a fatigue crack growth test gives a reasonable prediction of the abrasion rate. They has shown that the low strength of the BR material results in much smaller asperities being formed under steady state abrasion which results in a much slower abrasion rate. Conversely the strongest material NR has the longest tongue on the asperity and this in turn generates much larger values for the tearing energy at the tip of the asperity and this contributes to its poor abrasion resistance. Hong et al. (Hong et al., 2007) observed that BR compounds caused much slower wear than NR and SBR compounds. Arayapranee and Rempel (Arayapranee & Rempel, 2009) studied the cure characteristics, mechanical properties before and after heat ageing, and abrasion and ozone resistances of hydrogenated natural rubber (HNR), providing an ethylene–propylene alternating copolymer, vulcanizate and compared with those of natural rubber (NR), ethylene propylene diene terpolymer (EPDM) and 50:50 NR/EPDM vulcanizates. They reported that the highest abrasion resistance of the NR vulcanizate could be attributed to high unsaturated structure, as evident from its highest tensile strength compared to other vulcanizates. The abrasion loss of 48% HNR is higher than that of the NR vulcanizate, due to a reduction in the number of the double bonds. This suggests that abrasion resistance is heavily dependent on the unsaturation content in the backbone chain.

Index

Fillers

Fillers increase the stiffness of rubber in various degrees depending on quantity and quality of the fillers. The properties of rubber compounds are affected not only by the filler content but also by its structure and particle size. Despite outstanding resilience and high tensile strength, natural rubber possesses poor abrasion resistance. Thus, blending with high abrasion resistance rubbers and/or reinforcing by inorganic fillers are generally used to improve the abrasion resistance of NR and other rubbers. Carbon black and silica are two common fillers used to reinforce rubbers. However, high loadings of these fillers are required to obtain desirable properties. Incorporation of reinforcing fillers such as carbon black improves stiffness and strength of rubber (Tabsan et al., 2010). Hence, the abrasion resistance is improved by suppressing tearing of the rubber under the sliding contact (Gent

& Pulford, 1983). Arayapranee at el. studied the effect of filler type and loading on the abrasion loss (Arayapranee at el., 2005). They found that the incorporation of silica and carbon black reduces the abrasion loss of the natural rubber materials notably, whereas rice husk ash shows no effect with filler loading. Reinforcing fillers, silica and carbon black, interact preferentially with the natural rubber phase, as shown by the higher reduction of abrasion loss in the compounds. This improvement is probably due to the greater surface area and better filler-rubber interfacial adhesion resulting in an improved abrasion resistance. Fine particles actually reflect their greater interface between the filler and the rubber matrix and, hence, provide a better abrasion resistance than the coarse ones. Similar results were also reported by (Sae-oui et al., 2002).

Filled compounds are found to be less sensitive to the frictional force, whether wear took place by tearing or by smearing (Gent & Pulford, 1983). Carbon black is an additive with a decisive effect on the abrasion resistance. Hong et al. (Hong et al., 2007) investigated the effects of the particle size and structure of various carbon blacks on friction and abrasion behavior of filled natural rubber, styrene–butadiene rubber and polybutadiene rubber using a modified blade abrader. The effect of particle size and structure on abrasion resistance should be considered for the optimum design of desired wear properties. The worn surfaces of the rubber compounds filled with carbon black having smaller particle size and a more developed structure showed narrower spaced ridges and better abrasion resistance. It means that smaller particle size and better structure development of carbon black resulted in improved abrasion resistance. Yang et al. (Yang et al., 1991) reported that the abrasive wear of rubbers is strongly affected by the filler particles dispersed in the rubber matrix. The fillers are incorporated usually for the purposes of mechanical reinforcement and improving the conductivity of the neat resins. It is found that rigid filler particles normally increase the abrasive wear loss of the filled silicone rubbers. The wear rates of the filled silicone rubbers increase slowly with filler concentration until a critical volume fraction is reached, at which point they increase very rapidly with increasing filler. The critical filler fraction should carry important information, as it apparently divides two wear regimes dominated by different mechanisms. The first regime, where the filler concentration is low, is dominated by the properties of the neat resin. The increase of wear rate due to the filler is gradual here. In the cases of effective filler reinforcement, a reduction of wear rate can occur. The second regime is dominated by the filler's detrimental effects where the wear rate increases very rapidly with filler concentration. The stress concentration introduced by the rigid particles effectively creates a 'damage zone' surrounding the particles, a location where micro- cavitation and debonding takes place. Cavitation appears to dominate in the composites of very small filler particles, while debonding dominates when larger particles are involved. In view of the importance of carbon blacks on tread wear, it is surprising that relatively little understanding of the phenomenon has been set out in print. Although for synthetic rubbers such as BR and SBR it may seem unnecessary to look further than the dramatic enhancement of strength properties imparted by the use of particular grades of black, for NR such enhancement is modest and additional mechanisms for the effect of blacks on tread wear should be sought. In any case there is a consensus that high surface area, high surface activity and high structure promote tread wear resistance. Even so, the evidence that carbon black does not necessarily enhance the abrasion resistance of rubber under conditions of equal sliding suggests that the effect of carbon black on tread wear may in part be simply associated with stiffening, and hence reduced sliding, without weakening the compound as a high crosslink density would do. Arayapranee and Rempel (Arayapranee & Rempel, 2008) studied the effects of incorporation of three different fillers, i.e. rice husk ash (RHA), silica and calcium carbonate (CaCO3), over a loading range of 0-60 phr on the abrasion loss of 75:25 natural rubber (NR)/ethylene propylene diene monomer (EPDM) blends. The incorporation of silica reduced the abrasion loss of the 75:25 NR/EPDM blends notably, whereas CaCO3 showed a different trend in abrasion loss tending to increase it with an increase in CaCO3 loading. However, RHA showed less of an effect with filler loading. At a similar filler loading, silica filled 75:25 NR/EPDM blends had the lowest abrasion loss followed by RHA and CaCO3 filled 75:25 NR/EPDM blends.

Lubricants

Lubricants is widely used in the compounding of diene rubbers to improve the processability of the compounds and to impart the desired physicomechanical properties of rubber compounds and vulcanizates. The presence of a liquid can prevent moving surfaces from coming into inimate contact if viscous flow from the contact region is sufficiently sluggish. Lubricants, such as non-swelling fluids or dust, greatly reduce friction on smooth surface but the effect is smaller on rough surface. Changes in friction properties of rubber are possible by adding substantial amounts of standard lubricants, but this reduces strength, especially at high temperatures. Contrary, improvement in friction properties of rubbers based on blends NR and BR could be reached by introducing 0.5 wt% of K95 experimental lubricant (Jurkowska et al., 2006). Lubricant K95 added in a quantity of 0.5 wt% reduced the viscosity of rubber compound; it also improved compound flow in the mold. Mechanical properties of cured rubber not decrease while resistance to abrasion and fatigue increased. The influence of Lubricant K95 on reducing of the internal friction of rubbers is found.

Evstratov et al. (Evstratov et al., 1967) found that abrasion on a ribbed metal surface increases abruptly, by an order of magnitude or so, when the friction coefficient (□) exceeds about 1.4. Abrasion patterns were observed for □□above the critical value, but not for lower values. It did not matter whether □□was an unlubricated value for the compound or was determined by the presence of a lubricant. The renowned abrasion resistance of cis-BR compounds may relate to this observation; such compounds have low dry friction and form only very fine abrasion patterns. In spite of their low strength, their abrasion resistance can be excellent. When a lubricant is applied, a much finer pattern develops and the rate of abrasion is much lower.

Antioxidants

Gent and Pulford (Gent & Pulford, 1983) determined rates of wear have been determined for several rubber materials, using a razor-blade abrading apparatus at different levels of frictional power input, corresponding to different severities of wear, at both ambient temperature and at 100□C, and both in air and in an inert atmosphere. It is concluded that wear occurs as a result of two processes: local mechanical rupture (tearing) and general decomposition of the molecular network to a low-molecular-weight material (smearing). The decomposition process could, in principle, be ascribed to several mechanisms: thermal decomposition due to local heating during sliding; oxidative deterioration, possibly accelerated by local heating; and mechanical rupture of macromolecules to form reactive radical species. The most plausible mechanism of smearing appears to be oxidative consummation of scissions produced by mechanical stress, in much the same way as occurs during cold mastication of natural rubber. They provided rather convincing evidence of mechanochemical degradation of certain rubbers during abrasion by a razor blade. The degradation of carbon black-filled natural rubber (NR), styrene-butadiene rubber (SBR), and ethylene-propylene rubber (EPM) to a sticky material during blade abrasion occurred only in the presence of oxygen or thiophenol, but not in a nitrogen atmosphere (just as for cold mastication). Polybutadiene rubber (BR) produced only dry debris during abrasion, consist with the expectation that any free radicals of BR produced by main chain rupture would react with the polymer itself, leading to an increase in cross-linking rather than degradation. Carbon black-filled natural rubber, styrene-butadiene rubber, and ethylene-propylene rubber were particularly susceptible to decomposition and smearing, but for natural rubber and SBR the decomposition process was not observed in an inert atmosphere. It is attributed to molecular rupture under frictional forces followed by stabilization of the newly formed polymeric radicals by reaction with oxygen, if present, or with other polymer molecules, or with other macroradicals. Polybutadiene rubber produced only dry debris during abrasion. Radicals of BR produced by main chain rupture would react with the polymer itself, leading to an increase in cross-linking rather than degradation. Rates of wear have been found to increase with the applied frictional force raised to a power n. The value of n was between 2.5 and 3.5 for unfilled materials at ambient temperature. Filled materials were found to be less sensitive to the frictional force, whether wear took place by tearing or smearing, having values of the index n of 1.5-1.8. It is well known that for some conditions the surface of rubber becomes tacky during abrasion experiments, drum testing of tires and sometimes even for tires on the road. It has been suggested that either exudation of low molecular weight additives or degradation of the polymer to a material of low molecular weight could be responsible. Degradation might result from either thermal or mechanical stress, at high sliding speeds, such as skidding of a vehicle on locked wheels, frictional heating certainly causes degradation. However, the phenomenon of smearing is associated with conditions of mild abrasion, e.g. on smooth surfaces, and can occur even for low sliding speeds.

The most plausible mechanism of smearing appears to be oxidative consummation of scissions produced by mechanical stress, in much the same way as occurs during cold mastication of NR. Similar experimental observations to those of Gent and Pulford (Gent & Pulford, 1983) were previously obtained by Rudakov and Kuvshinski (Rudakov & Kuvshinski, 1967) for abrasion of NR and BR by a smooth indenter in air and in helium. They also gave a calculation suggesting that the rise in temperature of the rubber surface was quite inadequate to cause thermal degradation. However, this calculation ducks the possibility of local hotspots: the smaller the region of real contact, the higher is the calculated temperature rise, but we can only conjecture as to the size of the real contacts (Schallamach, 1967).

Schallamach (Schallamach, 1968) investigated the factors influencing smearing on the Akron laboratory abrader. He found that smearing could be prevented for NR tire tread compounds by carrying out abrasion in nitrogen or obviated by feeding a dust (magnesia proved most effective) into the nip between test piece and abrasive wheel. He concluded that oxidative degradation (to which he attributed smearing) affects the rate of abrasion in two distinct ways. If smearing occurs, the rate of abrasion is reduced (presumably because the "smear" acts as a lubricant). When the abrasion of a rubber is low in air, owing to smearing, its abrasion in nitrogen can become greater than in air. However, in air the less grossly degraded rubber is mechanically weakened, so that if smearing is obviated by the use of a suitable dust, the rate of abrasion is greater in air than in nitrogen. He also showed that the susceptibility of the compound to oxidative degradation can be influenced by the choice of antioxidant and other formulation details. Pulford (Pulford, 1983) studied antioxidant effects during abrasion of NR tire tread compounds by a razor blade. He reported that all compounds exhibited smearing at sufficiently low friction loads, but antioxidants reduce the critical frictional force below which smearing occurs. He found that antioxidants reduce the rate of wear for conditions in which smearing occurs but have no effect at higher severities. He considered this to be evidence of two mechanisms of wear, namely degradation at low frictional force and fracture at high frictional force. However, antioxidants also protect against fatigue crack growth, but only at low tearing energies (Lake, 1983). Thus it may not be necessary to invoke an entirely different mechanism of abrasion when smearing occurs. Instead, smearing can be seen as a complication superimposed on the general fracture mechanism of abrasion. Antioxidants can be used to, at least, partially restore the abrasion and crack growth resistance.

CONCLUSIONS

Abrasion resistance is the ability of a material to withstand mechanical action such as rubbing, scraping, or erosion that tends progressively to remove material from its surface. Such an ability helps to maintain the material's original appearance and structure. Numerous companies manufacture abrasion resistant products for a variety of applications, including products which can be custom fabricated to meet the needs of specific users. Abrasion resistant materials can be used for both moving and fixed parts. In vulcanized material or synthetic rubber compounds, a measure of abrasion resistance relative to a standard rubber compound under defined conditions. The properties of a vulcanized rubber can be significantly influenced by details of the compounding. Practical materials will have, in addition to the base polymer, fillers, antioxidants, crosslinking agents, accelerators etc. All of these can have an influence on the physical and chemical stability of the finished material. For example, rubber abrasion resistance can be related quantitatively to the frictional force and the crack growth characteristic of the rubber. Rigid filler particles normally increase the abrasive wear loss of the filled rubbers. A lubricant may cause a small decrease in frictional force but a dramatic decrease in abrasion. Antioxidants can be used to, at least, partially restore the abrasion and crack growth resistance because they are added to prevent ozone attack on the rubber, which can lead to the formation and growth of cracks.

REFERENCES

1. Arayapranee,W.; Na-Ranong, N. & Rempel, G. L. (2005). Application of Rice Husk Ash as Fillers in the Natural Rubber Industry. *Journal of Applied Polymer Science*, Vol. 98, pp. 34–41.
2. Arayapranee,W. & Rempel, G. L. (2008). A comparison of the properties of rice husk ash, silica, and calcium carbonate filled 75:25 NR/EPDM blends. *Journal of Applied Polymer Science*, Vol. 110, pp. 1165–1174.
3. Arayapranee,W. & Rempel, G. L. (2009). Synthesis and mechanical properties of diimide- hydrogenated natural rubber vulcanizates. *Journal of Applied Polymer Science*, Vol. 114, pp. 4066–4075.
4. Champ, D.H. ; Southern, E. & Thomas, A.G. (1974). Fracture mechanics applied to rubber abrasion. *American Chemical Society, Division of Organic Coatings and Plastics Chemistry*, Vol. 34, No. 1, pp. 237-243.
5. Chandrasekaran, M. & Batchelor, A.W. (1997). In situ observation of sliding wear tests of butyl rubber in the presence of lubricants in an X-ray micro focus instrument. *Wear*, Vol. 211(1), pp. 35–43.
6. Dick, J. S. (2001). Vulcanizate physical properties, performance characteristics and testing, In: *Rubber Technology Compounding and Testing for Performance*, 66-67, Hanser, ISBN 1-56990-278-X, Ohio, USA.

7. El-Tayeb, N. S. M. & Nasir, R. Md. (2007). Effect of soft carbon black on tribology of deproteinised and polyisoprene rubbers, *Wear*, Vol. 262, pp 350-361.
8. Evstratov, V. V.; Reznikovski, M. M.; Smirnova, L. A. & Sakhhinovabi, N. L. (1967). The mechanism of wear of tread rubbers, in: D.I. James (Ed.), Abrasion of Rubber, Maclaren, London, pp. 45–63.
9. Fukahori, Y. & Yamazaki, H. (1994). Mechanism of rubber abrasion—Part 1: abrasion pattern formation in natural rubber vulcanizate. *Wear*. Vol. 171, pp. 195–202.
10. Gent, A. N. (1989). A hypothetical mechanism for rubber abrasion. *Rubber Chemistry & Technology*, Vol. 62, pp. 750-756.
11. Gent, A. N. & Pulford, C. T. R. (1983). Mechanisms of rubber abrasion. *Journal of Applied Polymer Science*, Vol. 28, pp. 943-960.
12. Hong, C. K.; Kim, H. ; Ryu, C. ; Nah C. ; Huh, Y.-il & Kaang, S. (2007). Effect of paticle size and structure of carbon blacks on the abrasion of filled elastomer compounds, *Journal of Materials Science*, Vol 42, pp. 8391-8399.
13. Jurkowska, B.; Jurkowski, B.; Nadolny, K.; Krasnov, A. P.; Studniev, Y. N.; Pesetskii. S. S.; Koval, V. N.; Pinchuk, L. S. & Olkhov, Y. A. (2006). Influence of fluorine-containing lubricant on properties of NR/BR rubber. *European Polymer Journal*, Vol. 42, pp. 1676-1687.
14. Lake, G. J. (1983). Aspects of fatique and fracture of rubber. *Progress of Rubber Technology*, Vol. 45, pp. 89-143.
15. Liang, H.; Fukahori, Y.; Thomas, A. G. & Busfield, J. J. C. (2010). The steady state abrasion of rubber: Why are the weakest rubber compounds so good in abrasion? *Wear*, Vol. 268, pp. 756-762.
16. Muhr. A. H. & Roberts, A. D. (1992). Rubber abrasion and wear. *Wear*, Vol. 158, pp. 213-228.
17. Pandey, K. N.; Setua, D. K. & Mathur, G. N. (2003). Material behaviour fracture topography of rubber surfaces: an SEM study, *Polymer Testing*, Vol. 22, pp. 353–359.
18. Pulford, C. T. R. (1983). Antioxidant effects during blade abrasion of natural rubber. *Journal of Applied Polymer Science*, Vol. 28, pp. 709-713.
19. Ratner, S. B.; Farberova, I. I.; Radyukevich, O. V. & Lu're, E. G. (1967). Connection between wear resistance of plastics and other mechanical properties, In: *Abrasion of Rubber*, James, D. I. (Ed.), Macaren, London, pp. 145-154.
20. Roberts, A. D. (1988). Rubber adhesion at high rolling speeds. Journal of Natural Rubber Research. Vol 3, pp 239.

21. Rudakov, A. P. & Kuvshinski, E. V. (1967). Abrasion of rubber by a smooth indentor, In: *Abrasion of Rubber*, James, D. I (Ed.), Maclaren, London, pp. 36-44.
22. Sae-oui, P.; Rakdee, C. & Thanmathorn, P. (2002). Use of rice husk ash as filler in natural rubber vulcanizates: In comparison with other commercial fillers. *Journal of Applied Polymer Science*, Vol. 83, pp. 2485-2493.
23. Schallamach, A. (1957/58). Friction and abrasion, *Wear*, Vol. 1, pp. 384-417.
24. Schallamach, A. (1967). A note on the frictional temperature rise of tyres, *Journal of the Institution of the Rubber Industry*, Vol. 1, pp. 40-42.
25. Schallamach, A. (1968). Abrasion, fatigue, and smearing of rubber. *Journal of Applied Polymer Science*, Vol. 12, pp 281-293.
26. Tabsan, N.; Wirasate, S. & Suchiva, K. (2010). Abrasion behavior of layered silicate reinforced natural rubber. *Wear*, Vol. 269, pp. 394-404.
27. Thomas, A. G. (1958). Rupture of rubber. V. cut growth in natural rubber vulcanizates. *Journal of Polymer Science*. Vol. 31, pp. 467-480.
28. Thomas, A. G. (1974). Factors influencing the strength of rubbers. *Journal of Polymer Science*, Vol. 48. pp. 145-157.
29. Uchiyama, Y. & Ishino, Y. (1992). Pattern abrasion mechanism of rubber. *Wear*. Vol. 158, pp. 141–155.
30. Yang, A. C.-M.; Ayala, J. E. & Campbell Scott, J. (1991). Abrasive wear in filled elastomers. *Journal of Materials Science*, Vol. 26, pp. 5823-5837.

EFFECT OF COMPOUNDING INGREDIENTS ON ABRASION RESISTANCE

Rubber

In rubbery materials, when the smooth surface is abraded, periodic parallel ridged patterns are formed on the rubber surface. These typical patterns are held through all processes of rubber abrasion, on the surface of tires, conveyor belts, printing rolls and shoes for example, which are thus regarded as the essential basis of rubber abrasion. In the absence of any serious chemical decomposition the abrasion process initially results in the removal of small rubber particles just a few microns in size, leaving pits behind in the surface. With continued rubbing, larger pieces of rubber are removed. Although most weight loss is attributable to the larger pieces, it is thought that the detachment of the smaller particles initiates the abrasion process. The small particles have a characteristic size of 1-5 □m, but whether this relates to a structural unit in the rubber compound (Muhr & Roberts, 1992). Other suggestions are that mechanical rupture to produce the particles relates to flaws in the rubber, including dirt, or voids that cavitate leading to internal subsurface failure (Gent, 1989). A rolling experiment suggested that particle detachment might be linked to interfacial adhesion (Roberts, 1988). Schallamach (Schallamach, 1557/58, 1968) reported that rubber often develops a pattern of ridges perpendicular to the direction of abrasion. In the simplest case abrasion is produced by a line contact pulling a tongue of rubber from the ridge producing crack growth at the base of the tongue. Provided the surface configuration is in a steady state, the quantity of rubber abraded can be related quantitatively to the frictional force and the crack growth characteristic of the rubber. The abrasion of rubber results from mechanical failure due to excessively high local frictional stresses which are most likely to occur on rough tracks. Theories of abrasion thus require details of the local stresses, which together with the strength properties of the rubber may enable the rate of abrasion to be predicted. Gent and Pulford (Gent & Pulford, 1983) reported the reversal in the relative rates of wear of unfilled polybutadiene rubber comparing to those of unfilled natural rubber and styrene butadiene rubbers as frictional force increased. Fukahori and Yamazaki (Fukahori & Yamazaki, 1994) investigated the mechanism of the formation of the periodic ridges in rubber abrasion by designing the razor blade abrader. They reported that the driving force to generate the periodic surface patterns, and thus rubber abrasion consists of two kinds of periodic motions, stick-slip oscillation and the microvibration generated during frictional sliding of rubber. The stick-slip oscillation is the driving force to propagate cracks, then abrasion patterns and the microvibration with the natural frequency of the rubber induced in the slip phase of the stick-slip oscillation is another driving force for the initiation of the cracks. Although initial cracks originate in the slip region of the rubber surface, the propagation of the cracks is strongly excited in the stick region. Champ et al. (Champ et al., 1974) proposed the mechanism of rubber abrasion from a fracture mechanics point of view, relating the rate of wear to the crack growth resistance of the rubber. Although the concept of crack growth plays a very important role in abrasion, particularly in the growth of a single ridge, when consider that the essential subject of rubber abrasion. Liang et al. (Liang et al., 2010) investigated the blade abrasion of four different rubber materials, unfilled natural rubber, unfilled styrene butadiene rubber, unfilled polybutadiene rubber and carbon black filled styrene-butadiene rubber. Each is abraded until the steady state abrasion pattern is developed on the surface of moulded rubber wheels. The steady state conditions are measured using the weight loss per revolution of the wheel. The abraded surface is cut to examine the typical asperity profile. Each profile is modeled using finite element analysis to calculate the stored energy release rate for each combination of material and test condition. The stored energy release rate when combined with an independent measure of the rate of crack growth measured using a fatigue crack growth test gives a reasonable prediction of the abrasion rate. They has shown that the low strength of the BR material results in much smaller asperities being formed under steady state abrasion which results in a much slower abrasion rate. Conversely the strongest material NR has the longest tongue on the asperity and this in turn generates much larger values for the tearing energy at the tip of the asperity and this contributes to its poor abrasion resistance. Hong et al. (Hong et al., 2007) observed that BR compounds caused much slower wear than NR and SBR compounds. Arayapranee and Rempel (Arayapranee & Rempel, 2009) studied the cure characteristics, mechanical properties before and after heat ageing, and abrasion and ozone resistances of hydrogenated natural rubber (HNR), providing an ethylene–propylene alternating copolymer, vulcanizate and compared with those of natural rubber (NR), ethylene propylene diene terpolymer (EPDM) and 50:50 NR/EPDM vulcanizates. They reported that the highest abrasion resistance of the NR vulcanizate could be attributed to high unsaturated structure, as evident from its highest tensile strength compared to other vulcanizates. The abrasion loss of 48% HNR is higher than that of the NR vulcanizate, due to a reduction in the number of the double bonds. This suggests that abrasion resistance is heavily dependent on the unsaturation content in the backbone chain.

Index